Digital Waveform Processing and Recognition

Editor

C. H. Chen, Ph.D.

Professor and Chairman
Department of Electrical Engineering
Southeastern Massachusetts University
North Dartmouth, Massachusetts

CRC Press, Inc.
Boca Raton, Florida

Library of Congress Cataloging in Publication Data

Main entry under title:

Digital waveform processing and recognition

   Bibliography: p.
   Includes index.
   1. Signal processing—Digital techniques.
2. Pattern recognition systems. I. Chen, Chi-Hau,
1937-
TK5102.5D45   621.3819'598   80-29064
ISBN 0-8493-5777-2

   Direct all inquiries to CRC Press, Inc., 2000 Corporate Blvd., N.W., Boca Raton, Florida, 33431.

© 1982 by CRC Press, Inc.
Second Printing, 1983

International Standard Book Number 0-8493-5777-2
Library of Congress Card Number 80-29064

Printed in the United States

PREFACE

There is strong interest and numerous activity in digital signal processing and digital pattern recognition and their applications. Many industrial, academic, and government facilities are concerned with both processing and recognition of waveforms. A need clearly exists to integrate both digital signal processing and digital pattern recognition into a single volume book such that the common, as well as interrelated methodologies in *both* processing and recognition can be developed and applied to various applications.

The present volume is written with these objectives in mind. Chapters 2 and 3 are concerned with the basic principles of digital filtering and spectral analysis along with discrete detection and estimation, as well as some illustrative examples using speech and biomedical waveforms. Chapters 4 and 5 review the statistical and syntactic pattern recognition with application to signal processing. More detailed discussions on applications to speech, geophysics, sonar, and radar are presented in Chapters 6 to 10. Certain aspects of digital system implementation for waveform processing and recognition are considered in Chapter 11. The book should serve a dual purpose of reference and textbook in theory, design, and applications of digital waveform processing and recognition.

It is my pleasure to acknowledge the important contributions to this book by Professors K. S. Fu, D. G. Childers, and Alistair D. C. Holden. In addition, I express my appreciation to the Air Force Office of Scientific Research and the Office of Naval Research for their sponsorship of my research activities in pattern recognition and statistical signal processing.

C. H. Chen

THE EDITOR

Chi-Hau Chen received the Ph.D. degree in electrical engineering from Purdue University, West Lafayette, Ind., in 1965. From 1965 to 1968, he worked with ADCOM, Inc. and AVCO Systems Division, both in the greater Boston area, on various projects in digital communications and statistical data processing. Since 1968, he has been a member of the faculty of Southeastern Massachusetts University, North Dartmouth, where he is a Professor of Electrical Engineering. He teaches graduate courses in Digital Signal Processing, Pattern Recognition, Communication Theory, Speech Sonar and Seismic Signal Processing, Signal Detection Theory, and Time Series Analysis. His recent and current research interests have been in Statistical Pattern Recognition, Seismic Signal Processing and Discrimination, Discrete Orthogonal Transforms, Imagery Processing and Recognition, and Detection and Estimation Theory.

He has published over 100 technical papers in the areas of communications, pattern recognition, and signal processing. He is the author of the book, *Statistical Pattern Recognition,* Hayden Book Co., 1973; and editor of the books *Pattern Recognition and Artificial Intelligence,* Academic Press, Inc., 1976, and *Computer-Aided Seismic Analysis and Discrimination,* Elsevier Scientific Publishing Co., 1978. He was the director of the 1978 NATO Advanced Study Institute on Pattern Recognition and Signal Processing held in Paris, and editor of its Proceedings, published by Sijthoff & Noordhoff International in 1978.

Dr. Chen is a senior member of the Institute of Electrical and Electronic Engineers (IEEE), a member of the American Statistical Association, and Pattern Recognition Society.

CONTRIBUTORS

Donald G. Childers, Ph.D.
Professor of Electrical Engineering
University of Florida
Gainesville, Florida

K. S. Fu, Ph.D.
Goss Distinguished Professor of
  Engineering
Professor of Electrical Engineering
Purdue University
West Lafayette, Indiana

Alistair D. C. Holden, Ph.D.
Professor of Electrical Engineering and
  Computer Science
University of Washington
Seattle, Washington

To WANDA,
Ivan, and Christopher

TABLE OF CONTENTS

Chapter 1

INTRODUCTION

C. H. Chen

TABLE OF CONTENTS

I. DESCRIPTION OF DIGITAL WAVEFORMS

The measured data in many applications are a set of waveforms (or a waveform) from which desired information must be extracted. In the biomedical area, for example, the electrocardiogram (EKG) is taken from a patient and interpreted by a physician to determine whether the patient's heart is normal or abnormal. In the case of abnormality, further information may be obtained from the EKG. In practice, waveforms measured in many applications contain far more information than can be fully extracted by human users. Also, the large volume of data will make it very difficult if not impossible for human users to obtain the desired information in a reasonable amount of time. Computer-aided, interactive, and fully automatic techniques have been developed for processing and recognition. It is often necessary to digitize the waveforms before any processing and recognition can be performed by digital computers. The result of processing and recognition will be the output of the computer. For the EKG example, the processed result may be the spectral display, while the recognition result may be an interpretation of the EKG.

In this book the waveforms considered are in digital form. For processing, the input is thus a set of digital sequence or time-series data. For recognition, the input is usually some processed data such as the vector sample consisting of several feature measurements. Both digital systems for processing and recognition are also considered in this book.

II. PROCESSING OF DIGITAL WAVEFORMS

Processing includes practically all operations of the digital waveform by digital computers other than decision making and interpretations. Processing is closely related to recognition because a preliminary processing of waveforms is almost always needed for recognition. An important class of processing operations is the filtering of the data to minimize the effect of instrumentation noise or to constrain the data to certain frequency range. This can be done in both analog and digital form. Digital filtering has gained much acceptance because of its flexibility and high performance. Another important class of processing operations is the spectral analysis of waveforms. In addition to time-domain analysis of the data, frequency-domain analysis is essential in waveform study. Quite often both time-domain and frequency-domain processing are needed to derive certain desired information.

III. RECOGNITION OF DIGITAL WAVEFORMS

The main recognition objective is to classify a pattern or, in this case, the digital waveform into one of several possible categories (classes). Another recognition objective may be to obtain certain descriptions or structures of the data. However, a key problem in recognition is the feature extraction, i.e., the determination of characteristic features that can discriminate data from different classes and correctly identify samples from the same classes. By considering the patterns as statistical in nature, the statistical pattern recognition deals with the statistical description of digital signals, the extraction of mathematical features, decision rules, clustering, and the estimation of parameters and densities. Syntactic pattern recognition, on the other hand, deals with primitive selection and pattern grammar, syntactic classification and error correcting, and parsing and syntactic clustering.

IV. OVERVIEW OF THE BOOK

There is now very strong interest and numerous developments in both the digital signal processing and digital pattern recognition fields and their applications. By considering both fields in the same book, the common, interrelated methodologies in both processing and recognition can be developed and applied to various applications.

The fundamentals of digital signal processing considered in Chapter 2 deal with the specific topics of nonrecursive digital filters and discrete estimation and detection. The latter, which is not discussed in digital signal processing texts, is of fundamental importance to the processing and classification of digital waveforms. Chapter 3 is concerned primarily with the spectral analysis, with illustrative examples on speech and biomedical waveforms. Chapter 4 provides a comprehensive treatment of the statistical theory in pattern recognition. An excellent presentation of the syntactic approach to pattern recognition with application to signal processing is provided in Chapter 5. Chapter 6 examines several important techniques in speech processing. A very detailed presentation on geophysical data in both processing and recognition is given in Chapters 7 and 8. Both teleseismic and intrusion-detection seismic data are examined in detail as illustrative examples. Listings of several important computer programs are also provided. Key aspects of sonar and radar signal processing are examined in Chapter 9 and 10, respectively. Finally in Chapter 11, we consider the digital system implementation problems for both processing and recognition.

BIBLIOGRAPHIC NOTE

In each of the two fields, digital signal processing and pattern recognition, there are now well over 20 books published in the last 15 years. The number of articles published far exceeds 1000 in each field. The following short bibliography is representative of the enormous amount of published literature in both fields.

1. *Proc. Int. Conf. on Pattern Recognition,* Institute of Electrical and Electronics Engineers Computer Society, New York, 1973, 1974, 1976, 1978, 1980.
2. *Proc. IEEE Int. Conf. on Acoustics, Speech and Signal Processing,* Institute of Electrical and Electronics Engineers, New York, 1976, 1977, 1978, 1979, 1980.
3. Oppenheim, A. V. and Schafter, R. W., *Digital Signal Processing,* Prentice-Hall, Englewood Cliffs, N. J., 1975.
4. Oppenheim, A. V., Ed., *Applications of Digital Signal Processing,* Prentice-Hall, Englewood Cliffs, N.J., 1978.
5. Peled, A. and Liu, B., *Digital Signal Processing, Theory, Design and Implementation,* John Wiley & Sons, New York, 1976.
6. Fu, K. S., *Syntactic Methods in Pattern Recognition,* Academic Press, New York, 1974.
7. Fu, K. S., Ed., *Applications of Pattern Recognition,* CRC Press, Boca Raton, Fla., in press.
8. Chen, C. H., *Statistical Pattern Recognition,* Hayden Book, Rochelle Park, N.J., 1973.
9. Chen, C. H., Ed., *Pattern Recognition and Signal Processing,* Sijthoff & Noordhoff, The Netherlands, 1978.

Chapter 2

FUNDAMENTALS OF DIGITAL SIGNAL PROCESSING

C. H. Chen

TABLE OF CONTENTS

I. INTRODUCTION

The fundamentals of digital signal processing encompass a broad spectrum of topics: sampling and reconstruction, transform methods, digital filtering, spectral analysis, adaptive signal processing, discrete estimation and detection, and signal processing hardware and software, etc. These topics are examined on a selective basis in this chapter and in Chapters 3, 7, 8, and 11. The basic principles of digital signal processing are well described in introductory level books[1,2] and more advanced level books.[3,4] These books are recommended to the reader for a complete introduction to the subject. Following an overview of the field of digital signal processing, this chapter will deal with several aspects of the field.

A typical digital signal processor starts with the digitization (i.e., sampling) of a continuous waveform at a sampling rate that is at least twice the highest significant frequency of the waveform. The remaining part of this analog-to-digital conversion is quantization of the discrete data, which, unlike sampling, is an irreversible process. The resulting data are processed digitally by filtering, transforming, and the combination of various techniques so that the desired information can be extracted or enhanced. The output of the digital processor can be in digital form or, in many applications, converted to analog form. Both the processed digital and analog data can be displayed as needed.

II. OVERVIEW OF DIGITAL SIGNAL PROCESSING

The field of digital signal processing has grown enormously in the past 15 years to provide firm theoretical background for a number of topics, as mentioned in the previous section. The major subdivisions of the field are digital filtering and spectrum analysis. The field of digital filtering is further divided into nonrecursive or finite impulse response (FIR) filters and recursive or infinite impulse response (IIR) filters. The latter may be considered as discrete counterparts of the continuous linear time invariant system. The field of spectrum analysis is broken into calculation of spectra via the discrete Fourier transform (DFT) and via statistical techniques as in the case of random signals, e.g., quantization noise in a digital system. The fast Fourier transform (FFT), which is a computationally efficient procedure to calculate DFT, and the related area of fast convolution are almost exclusively used in practical spectrum analysis techniques. The remaining aspects of digital signal processing are the important topics of implementation of digital systems and application areas. The applications are treated in detail in this book and the book by Oppenheim.[5] A good understanding of the issues involved in practical implementation of digital systems for signal processing in finite precision software or hardware, is essential to make good use of the theoretical study. The flexibility and greater accuracy offered by digital processing will help motivate the development of new digital components. Eventually, digital signal processing will replace analog processing in most applications, as the digital computer replaces the analog computer.

III. DESIGN OF NONRECURSIVE DIFFERENTIATOR AND HILBERT TRANSFORMER

In this section, we consider a simple technique of designing nonrecursive digital filters, as proposed by Gold and Radar.[6] The technique is flexible enough so that only a minor change in computer program statements is required to change from the differentiator to the Hilbert transformer or vice versa.[7] The effects of various window functions and the size of the window on the frequency response of the resulting filter are also examined.

The transfer functions of an ideal differentiator and an ideal Hilbert transformer are given by

$$F_D\,(j\omega) \;=\; j\omega, \qquad -\pi < \omega T < \pi \tag{1}$$

and

$$F_{H\,(T)} \;=\; \begin{cases} +j & 0 < \omega T < \pi \\[2mm] -j & -\pi < \omega T < 0 \end{cases} \tag{2}$$

respectively. Here, T is the sampling period. The real part of frequency response is zero for each filter. The subscripts D and H refer to the differentiator and the Hilbert transformer, respectively. The impulse response β_n is the Fourier coefficient defined by[6]

$$\beta_n \;=\; \frac{1}{2\pi} \int_{-\pi}^{\pi} F(\omega T)e^{+j\omega Tn}d(\omega T) \tag{3}$$

For the differentiator

$$\beta_{nD} \;=\; \begin{cases} \dfrac{(-1)^n}{n} & ,\; n \neq 0 \\[3mm] 0 & ,\; n = 0 \end{cases} \tag{4}$$

For the Hilbert transformer

$$\beta_{nH} \;=\; \begin{cases} -\dfrac{2}{n\pi} & ,\; n \neq 0 \text{ and } n \text{ odd integer} \\[3mm] 0 & ,\; n = 0 \text{ or } n \text{ even integer} \end{cases} \tag{5}$$

As is well known, weighting functions (also called window functions) are used to modify and truncate a Fourier series that represents a periodic function. The resulting Fourier representation, which exhibits Gibb's phenomenon, is an approximation to the original function.

Let $\omega(n)$ be the window function and N, a power of 2, be the size of the window. The window functions typically employed are

(1) Rectangular window

$$\omega\,(n) \;=\; \begin{cases} 1, & |n| < N \\[2mm] 0, & |n| \geqslant N \end{cases} \tag{6}$$

(2) Triangular window

$$\omega\,(n) \;=\; \begin{cases} 1 - \dfrac{|n|}{|N|} & ,\; |n| < N \\[3mm] 0 & ,\; |n| \geqslant N \end{cases} \tag{7}$$

(3) Parabolic window

$$\omega(n) = \begin{cases} 1 - \left|\dfrac{n}{N}\right|^2, & |n| < N \\ \\ 0, & |n| \geqslant N \end{cases} \qquad (8)$$

(4) Unraised half-cosine window

$$\omega(n) = \begin{cases} \cos\dfrac{n\pi}{2N}, & |n| < N \\ \\ 0, & |n| \geqslant N \end{cases} \qquad (9)$$

(5) Hanning window

$$\omega(n) = \begin{cases} 0.5 + 0.5\cos\dfrac{n\pi}{N}, & |n| < N \\ \\ 0, & |n| \geqslant N \end{cases} \qquad (10)$$

(6) Blackman's window

$$\omega(n) = \begin{cases} 0.42 + 0.5\cos\dfrac{n\pi}{N} + 0.08\dfrac{2n\pi}{N} & |n| < N \\ \\ 0, & |n| \geqslant N \end{cases} \qquad (11)$$

Computationally, β_n is calculated from Equation 3 by using FFT and then multiplied by the window function. Then, we take the FFT of $\beta_n \omega_n$ to obtain the desired frequency response of the nonrecursive digital filter. The imaginary parts, $\text{Im}[F(j\omega)]$, of the frequency response of the differentiator and the Hilbert transformer are shown in Figures 1 and 2, respectively. $\text{Im}[F_D(j\omega)]$ is continuous at $\omega T = 0$ but possesses a jump of $2\pi/T$ at $\omega T = \pi$. $\text{Im}[F_H(j\omega)]$ exhibits a discontinuity at $\omega T = 0$ as well as $\omega T = \pi$. The positive n parts of the impulse response β_n of the filters are shown in Figures 3 and 4 with a total number of points $N_t = 256$. For the rectangular window and $N = 16$, the frequency responses of the truncated impulse response are shown in Figures 5 and 6, respectively, for the differentiator and the Hilbert transformer. For $N = 8$, the corresponding results are shown in Figures 7 and 8. The number of ripples is doubled from $N = 8$ to $N = 16$ but $N = 16$ provides a much better approximation to the ideal frequency response. The frequency responses corresponding to the triangular window are shown in Figures 9 and 10 for $N = 6$ and in Figures 11 and 12 for $N = 8$. It is noted that the ripples are considerably reduced in going from rectangular to triangular windows. Again, $N = 16$ provides much better frequency responses compared to $N = 8$. By changing the triangular to other windows, only slight improvement in frequency response is obtained (see Figures 13 to 16 and 17 to 20. The Blackman's window provides the most smooth frequency response in comparison with the parabolic, half-cosine, and Hanning windows, but it has the largest transition bandwidth. It may be concluded that each window function has its own advantages and disadvantages. The choice of a particular window depends on the requirements of the filter used.

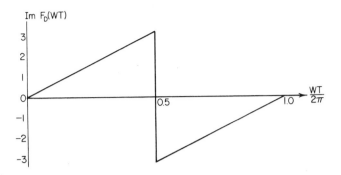

FIGURE 1. Ideal frequency response $F_D(WT)$, imaginary part.

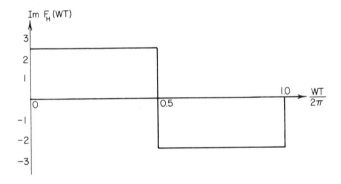

FIGURE 2. Ideal frequency response, $F_H(WT)$ for Hilbert transformer, imaginary part.

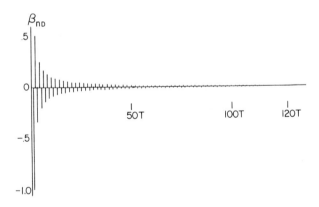

FIGURE 3. Ideal β_{nd}, positive n part. $N_+ = 256$

Although only the differentiator and the Hilbert transformer are discussed in this section, the method of approach can be used just as easily for other nonrecursive filter designs.

IV. DIGITAL CROSS-CORRELATOR

Digital cross-correlation and matched filtering operations are frequently used in digital communications, sonar and radar signal processing, etc. Effectiveness of these

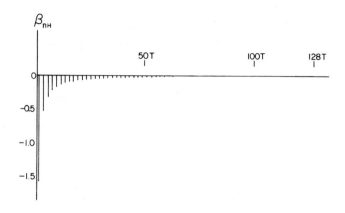

FIGURE 4.   Ideal β_{nh}, positive n part, $N_+ = 256$.

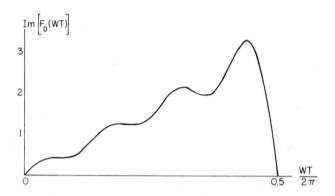

FIGURE 5. $F_D(WT)$, rectangular window, $N = 16$.

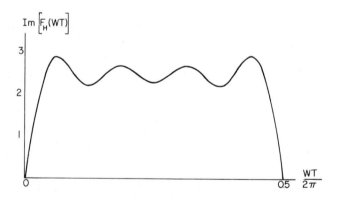

FIGURE 6. $F_H(WT)$, rectangular window, $N = 16$.

operations depends highly on the noise and quantization. Figure 21 is a block diagram of a digital correlator. The input waveform $x(t) = s(t) + n(t) + i(t)$ consists of a signal $s(t)$, random noise $n(t)$, and interference $i(t)$. The reference waveform $r(t)$ can be deterministic. These waveforms are sampled once every T seconds and then quantized, usually with a symmetric quantizer. The output variables a_i, which is random,

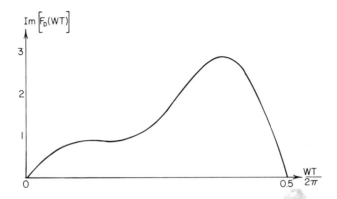

FIGURE 7. $F_D(WT)$, rectangular window, N = 8.

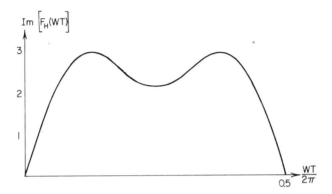

FIGURE 8. $F_H(WT)$, rectangular window, N = 8.

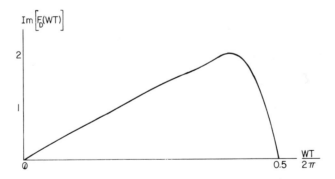

FIGURE 9. $F_D(WT)$, triangular window, N = 16.

and b_i, which is deterministic, are multiplied and summed to form a decision variable d as the correlator output,

$$d = \sum_{i=1}^{N} a_i b_i \qquad (12)$$

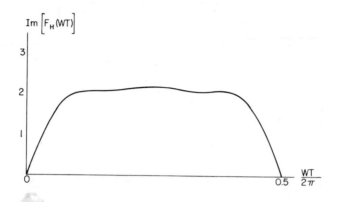

FIGURE 10. F_H(WT), triangular window, $N = 16$.

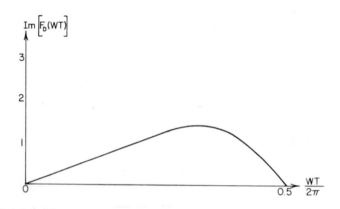

FIGURE 11. F_D(WT), triangular window, $N = 8$.

where N is the total number of samples taken in each waveform over a given duration. There are many interpretations of the inner product in the form of Equation 12 in the areas of power spectral estimation, numerical operations such as integration and interpolation, digital correlation detection, and nonrecursive digital filtering. Typically, the elements of the Gaussian random sequence of Equation 12 are

$$x_i = A \sin (2\pi i T) + n (iT)$$

$$i = 0,1, \cdots, N - 1 \tag{13a}$$

in which n(t) is a sample function of a stationary zero mean Gaussian random process of variance σ_N^2. Here, i(t) is assumed as zero. The SNR (signal-to-noise ratio) is $A^2/2\sigma_n^2$. The constant reference values are

$$r_i = \sin (2\pi i T)$$

$$i = 0,1, \cdots, N - 1 \tag{13b}$$

When the reference waveform is the same as the signal s(t), as illustrated above, the correlator is a digital-matched filter.

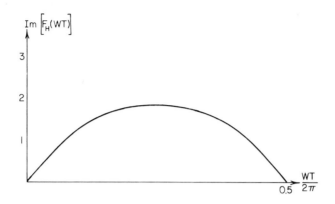

FIGURE 12. F_H(WT), triangular window, N = 8.

Let the overbar denote the expectation. The statistics of d are given by

$$\overline{d} = \sum_{i=1}^{N} \overline{a_i} b_i$$

$$\sigma_d^2 = \sum_{i=1}^{N} \sum_{i=1}^{N} \overline{(a_i - \overline{a_i})(a_j - \overline{a_j})} \; b_i b_j$$

$$= \sum_{i=1}^{N} \sum_{j=1}^{N} \alpha_{ij} b_i b_j + \sum_{i=1}^{N} [\overline{a_i^2} - (\overline{a_i})^2] b_i^2$$

$$= 2 \sum_{k=1}^{N} \sum_{\ell=1}^{N-k} \alpha_{\ell, \ell+K} b_\ell b_{\ell+k} + \sum_{i=1}^{N} [\overline{a_i^2} - (\overline{a_i})^2] b_i^2 \qquad (14)$$

where the correlation α_{ij} forms a covariance matrix that is determined by the noise statistics. Define a figure of merit F as a measure of separation of the probability density functions with and without signal as

$$F = \frac{\overline{d}_{s+N} - \overline{d}_N}{\sigma_N} \qquad (15)$$

which will be useful to estimate the performance degradation of the digital correlation, with respect to the analog correlation due to the noise and quantization. By using the efficient numerical method of Kellogg[8] for computing the autocorrelation function of the quantizer output, Tufts et al.[9,10] computed the figure of merit F as a function of SNR for various interference-to-noise ratios, and in the case of zero interference, the decrease of F for different noise correlation conditions. For the noise correlation of exp (−3t) and SNR = 64, the decrease of $20 \log_{10} F$ is 6.419 where x_i and r_i in Equation 13 are quantized into four and two levels, respectively.[10] A much larger quantization number obviously is needed for each quantizer in practice. For a given probability density function such as Gaussian, it is possible to determine the detection and false

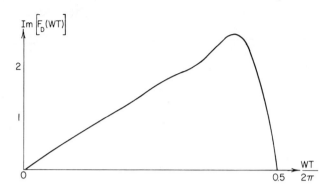

FIGURE 13. $F_D(WT)$, parabolic window, $N = 16$.

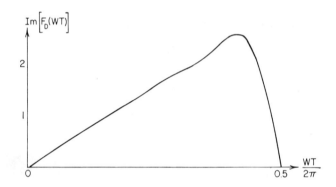

FIGURE 14. $F_D(WT)$, unraised half-cosine window, $N = 16$.

alarm probabilities for the correlator. Such probabilities depend on $\bar{d}_{s+N} - \bar{d}_N$ and thus on F. The exact change on such probabilities due to noise and quantization, however, has not been examined.

V. DISCRETE DETECTION OF KNOWN SIGNALS

In this and following sections, we shall consider the problem of discrete detection and estimation, which is of fundamental importance in digital signal processing. For notational clarity, vectors are denoted by boldface lowercase letters such as m, x, and n, and matrices are denoted by boldface uppercase letters such as Q, F, and G. All vectors are column vectors. The asterisk is used to denote complex conjugate transposition so that x* is a row vector whose elements are the complex conjugate of the elements of x. Generally, random vectors are assumed to be complex. A complex random vector $z = \xi + i\eta$ must possess the following properties, which are invariant under linear transformation:

1. The real part ξ and the imaginary part η are both real random vectors with the same covariance matrix.
2. For all components j and k, $E(\xi_j \eta_k) = -E(\xi_k \eta_j)$.

Further information on complex random variables is available.[11,12]

Now the binary detection problem is one of deciding whether or not a signal is

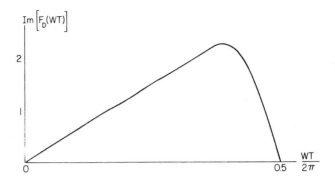

FIGURE 15. $F_D(WT)$, Hanning window, $N = 16$.

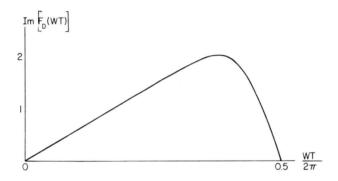

FIGURE 16. $F_D(WT)$, Blackman's window, $N = 16$.

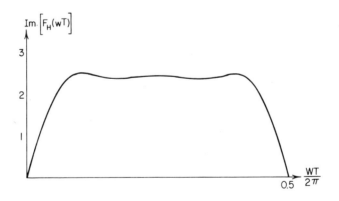

FIGURE 17. $F_H(WT)$, parabolic window, $N = 16$.

present. A choice is made between the following hypotheses:

$$H_0 : x = n$$

$$H_1 : x = Ms + n \qquad (16)$$

where x is the received vector, s and n are signal and noise vectors, respectively, and M is a known transformation matrix (not necessarily square). In the known signal case

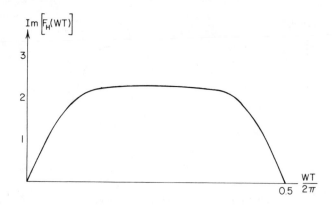

FIGURE 18. $F_H(WT)$, Hanning window, N = 16.

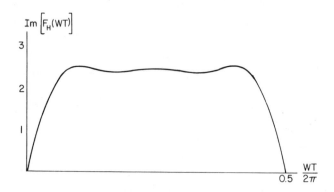

FIGURE 19. $F_H(WT)$, unraised half-cosine window, N = 16.

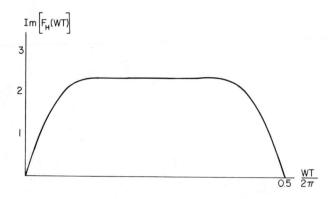

FIGURE 20. $F_H(WT)$, Blackman's window, N = 16.

the optimum procedure for the Bayes, Neyman-Pearson, and minimax decision criteria is a likelihood ratio test in wbich the value of the likelihood ratio

$$L(x) = \frac{p(x|H_1)}{p(x|H_0)}$$

is compared with a threshold value L_o, and the decision "signal present" is made when

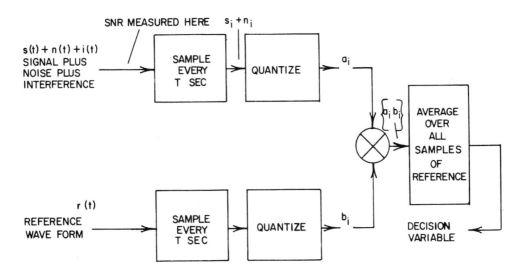

FIGURE 21. Block diagram of a digital correlator. (From Tufts, D. W., Knight, W., and Rorabacher, D., *Proc. IEEE (Letters),* 56, 79, 1969. With permission.)

the threshold is exceeded. The level of the threshold depends on the particular criterion used.[13]

When the noise is Gaussian, the likelihood ratio takes the form

$$L(x) = \frac{\exp\left\{-\tfrac{1}{2}\lambda\,(x-Ms)^*Q^{-1}\,(x-Ms)\right\}}{\exp\left\{-\tfrac{1}{2}\lambda\,(x^*Q^{-1}x)\right\}}$$

where $\lambda = 2$ for complex x and $\lambda = 1$ for real x. Also, Q is the noise covariance matrix. The likelihood ratio may be reduced to

$$L(x) = \exp\left\{-\frac{1}{2}\,[s^*M^*Q^{-1}\,Ms - s^*M^*Q^{-1}x - x^*Q^{-1}Ms]\right\}$$

The first term does not depend on the particular observation or received x. Taking logarithms, we are led to decide "signal present" whenever the real scalar test statistic

$$y = \frac{1}{2}\,(s^*M^*Q^{-1}\,x + x^*Q^{-1}\,Ms) = Re\,(s^*M^*Q^{-1}x) \qquad (17)$$

exceeds a suitably chosen threshold. The factor λ is incorporated into the setting of the threshold level.

The distribution of the test statistic is Gaussian with means

$$E\,(y|H_0) = 0$$

and

$$E\,(y|H_1) = s^*\,M\,Q^{-1}\,Ms$$

under H_0 and H_1, respectively. The variance is the same under either hypothesis and is given by var $(y) = S^*\,M^*Q^{-1}\,Ms/\lambda$.

When n is non-Gaussian and only second order statistics are known, one frequently seeks a linear transformation of the form

$$y = \frac{1}{2}[k*n+x*k] = Re\,(k*x) \tag{18}$$

such that the output signal-to-noise ratio is maximized. The quantity

$$r_o = \frac{\text{change in mean-squared output due to signal}}{\text{mean-squared output for noise only}}$$

or

$$r_o = \frac{E(y^2\,|H_1) - E\,(y^2\,|H_0)}{E(y^2\,|H_0)} \tag{19}$$

is usually called the output signal-to-noise ratio. The maximum of r_o can be shown to occur when [11]

$$k* = s*M*\,Q^{-1} \tag{20}$$

Substituting the k* of Equation 20 into Equation 18 yields the same as Equation 17 for the test statistic.

A slightly different approach is to maximize the detection index defined as

$$d = \frac{\text{change in average output level due to signal}}{\text{standard deviation of output}}$$

or

$$d = \frac{E(y\,|H_1) - E(y\,|H_0)}{\sqrt{E(y^2\,|H_0) - E^2\,(y\,|H_0)}} \tag{21}$$

It can be shown[11] that minimizing d also leads to the same test statistics given by Equation 17.

VI. DISCRETE ESTIMATION OF RANDOM SIGNAL

If the signal vector in the expression $x = MS + n$ is random, then the problem arises of estimating s given a particular observation x. Let the mean vector and covariance matrix of s be u and P, respectively. It is well-known that the best estimate of s, given x for a quadratic cost criterion, is the conditional mean $E(s/x)$.

If s and n are both Gaussian and independent, then x and s are jointly Guassian. Then, the conditional mean is also the best estimate for a variety of other cost criteria. To obtain the conditional mean $E(s/x)$, we consider the parameters of the distribution of the vector formed by s and x:

$$E\begin{bmatrix} s \\ x \end{bmatrix} = \begin{bmatrix} u \\ Mu \end{bmatrix} \quad \text{and} \quad cov\begin{bmatrix} s \\ x \end{bmatrix} = \begin{bmatrix} P & P\,M* \\ MP & MPM*+Q \end{bmatrix}$$

It can be shown that[11]

$$\hat{s} = E(s|x) = u + PM^* [MPM^* + Q]^{-1} (x - Mu) \qquad (22a)$$

or

$$\hat{s} = u + [P^{-1} + M^*Q^{-1}M]^{-1} M^*Q^{-1} (x-Mu) \qquad (22b)$$

and

$$cov (s|x) = P - P M^* [MPM^* + Q]^{-1} MP \qquad (23a)$$

$$= [P^{-1} + M^*Q^{-1} M]^{-1} \qquad (23b)$$

Note that the covariance does not depend on the observed value of x.

When s and n are nonGaussian, the conditional mean is still the best estimate for a quadratic cost criterion. However, sometimes convenient expressions for computing the conditional mean do not exist, and frequently, available information is limited to second order statistics.

One procedure is to seek the linear estimate that minimizes the mean-square error. That is, one seeks an estimate of the form

$$\hat{s} = a + Bx \qquad (24)$$

and determines a and then B to minimize

$$e = \text{trace} \{E[(s - \hat{s}) (s - \hat{s})^*] \}$$

$$= E[(s - \hat{s}) * (s - \hat{s})] \qquad (25)$$

With some algebraic manipulations,[11] it can be shown that $a = [I - BM]u$ and $B = PM^*[MPM^* + Q]^{-1}$ and thus the estimate is the same as Equation (22a). Furthermore, the error covariance is the same as Equation 23a.

The estimation of unknown nonrandom signal parameter, detection of random signal, and detection of signal with unknown nonrandom parameters, etc., are discussed by Cox.[11]

VII. ARRAY PROCESSOR AND MULTICHANNEL FILTERING

The method of approach in the last two sections can be extended to an array of, say, L sensors located in some region of space.[14] The waveform $x_j(t)$ at the output of the jth sensor is related to the signal of interest s(t) by an equation of the following form when the signal is present:

$$x_j (t) = \int m_j (t - v) s(v) \, dv + n_j (t) \qquad (26)$$

where $m_j(t)$ is the linear transformation corresponding to the impulse response of the transmission medium leading to the jth sensor. In vector notation, the detection problem is to choose between the following hypotheses:

$$H_0: x(t) = n(t)$$

$$H_1: x(t) = \int m(t - v)\, s(v)dv + n(t) \tag{27}$$

The array processor must then choose a set of filters $\{h_1(t), h_2(t) ---, h_L(t)\}$ which maximizes the detection index. The array processor output is

$$y(t) = \sum_{i=1}^{L} \int x_i(v)\, h_i(t-v)\, dv \tag{28}$$

The optimum array processor selects the filter set such that prewhitening and matched filtering operations are performed.

Another approach to the problem is the adaptive multichannel filtering.[15] A filter of temporal length K coefficients per channel, f_k, ℓ; $k = 1, ---, K$, $\ell = 1, 2, ---, L$, convolves with the incoming signal to give rise to the filtered output

$$y_t = \sum_{k=1}^{K} \sum_{\ell=1}^{L} f_{k,\ell}\, x_{t-k,\ell} \tag{29}$$

where t is the time index. The filter coefficients can be chosen by minimizing some criterion function. The adaptive digital filtering algorithms and the stochastic approximation algorithms may be used to determine the filter coefficients.

REFERENCES

1. **Stearns, S. D.**, *Digital Signal Analysis*, Hayden Book, Rochelle Park, N.J., 1975.
2. **Peled, A. and Liu, B.**, *Digital Signal Processing, Theory, Design and Implementation*, John Wiley & Sons, New York, 1976.
3. **Oppenheim, A. V. and Schafer, R. W.**, *Digital Signal Processing*, Prentice-Hall, Englewood Cliffs, N.J., 1975.
4. **Rabiner, L. R. and Gold, B.**, *Theory and Application of Digital Signal Processing*, Prentice-Hall, Englewood Cliffs, N.J., 1975.
5. **Oppenheim, A. V.**, *Applications of Digital Signal Processing*, Prentice-Hall, Englewood Cliffs, N.J., 1978.
6. **Gold, B. and Rader, C. M.**, *Digital Processing of Signals*, McGraw-Hill, New York, 1969.
7. **Chen, C. H.**, A Note on the Design of Nonrecursive Differentiator and Hilbert Transformer, Tech. Rep. TR-EE-73-3, Southeastern Massachusetts University, North Dartmouth, July, 1973.
8. **Kellogg, W. C.**, Information rates in sampling and quantization, *IEEE Trans. Inf. Theory*, IT-13, 506, 1967.
9. **Tufts, D. W., Knight, W., and Rorabacher, D.**, Effects of quantization and sampling in digital correlators and in power spectral estimation, *Proc. IEEE (Letters)*, 56, 79, 1969.
10. **Tufts, D. W. and Knight, W.**, Statistics of the inner product of a quantized, random vector and a constant vector: effects of component correlation, *Proc. IEEE (Letters)*, 56, 821, 1969.
11. **Cox, H.**, Interrelated Problems in Estimation and Detection I and II, *Proc. NATO Advanced Study Inst. Signal Processing with Emphasis on Underwater Acoustics*, Enschede, Netherlands, August, 1968.

12. **Goodman, N. R.**, Statistical analysis based on a certain multivariate complex Gaussian distribution, *Ann. Math. Stat.*, 34, 152, 1963.
13. **Helstrom, C. W.**, *Statistical Theory of Signal Detection*, 2nd ed., Pergamon Press, Elmsford, N.Y., 1968.
14. **Cox, H.**, Optimum arrays and the Schwartz inequality, *J. Acoust. Soc. Am.*, 45, 228, 1969.
15. **Chang, C. Y.**, Adaptive multichannel filtering, *Proc. ICASSP*, 80, 462, 1980.

Chapter 3

DIGITAL SPECTRAL ANALYSIS

D. G. Childers

TABLE OF CONTENTS

I. INTRODUCTION

Researchers in nearly all areas within electrical engineering are interested in some aspect of spectral analysis. In the past, spectral analysis and signal processing have often been considered as separate analysis tools for disassembling the data. Spectral analysis, synonymous with harmonic analysis, was used to translate a signal from the time domain into the frequency domain where, hopefully, certain features of the signal could be more easily discerned, e.g., signal bandwidth. Various algorithms for performing spectral analysis arose over the years, but almost never were those algorithms related to signal processing procedures, which were usually considered distinct analysis techniques used to study temporal, e.g., transient, phenomena of the data.

In recent years, spectral analysis and signal processing have become interwoven. Basically, the common ground is data modeling. The objective of the analyst is to develop an accurate model of the available time series data. This model then has certain spectral properties and is usually parametric, i.e., represented by a set of parameters. The model also has a time domain function that makes a prediction or calculates a moving average. This duality of properties between the frequency domain and the time domain is certainly not new, but the thread of data modeling that weaves the properties of the two domains together is a relatively new idea in engineering, brought to us via mathematics and, especially, statistics.

In this chapter, we briefly review the "older" spectral analysis procedures, e.g., Blackman-Tukey and the periodogram, describe the "newer" methods of Bartlett and Welch, and develop the more modern procedures known collectively as linear prediction, autoregression, and maximum entropy. Along the way we shall discuss the maximum likelihood method and show how the various methods are related to mean-square estimation.

The various techniques shall be illustrated by analyzing a simulated data set, a speech waveform, and some electroencephalogram (EEG) data. We shall also show how linear prediction coefficients can be used in a *pattern recognition* problem such as speaker recognition. Further, an example of inverse filtering will be given. And finally, we shall show how linear prediction can be applied to the EEG as a technique for noise (or background) reduction.

In summary, we intend to illustrate that data modeling procedures are powerful general analysis techniques that can be used to study the properties of data in either the frequency or time domain, and further, they can be and have been used in the field of pattern recognition to extract features of the signal, which in turn can be used to classify the data.

Many of the ideas discussed here are being extended to image processing and multichannel data analysis, but the subject is too extensive for inclusion in this chapter. Rather, we shall focus on the analysis of time series.

II. OVERVIEW

In the sections to follow we shall find it convenient to share some common terminology and notation. First, we shall represent a sampled time series as either

$$x(nT) = x(n) = x_n \tag{1}$$

where T is the sampling interval. The one-sided z-transform of this series will be given by

$$X(z) = \sum_{0}^{\infty} x_n z^{-n} \tag{2}$$

The "spectrum" is found by letting $z = e^{j\omega T}$ and we frequently suppress T or assume $T = 1$. Thus

$$X(e^{j\omega T}) = X(\omega) = X(z)|_{z = e^{j\omega T}} \tag{3}$$

The three basic data models we shall consider are autoregression (AR), moving average (MA), and autoregressive-moving average (ARMA).

The autoregressive model may be considered to produce data y_n (or an output) when excited by a white noise sequence, x_n, as follows

$$y_n = a_1 y_{n-1} + a_2 y_{n-2} + \ldots + a_n y_{n-N} + b_0 x_n$$

$$= \sum_{1}^{N} a_i y_{n-i} + b_0 x_n \tag{4}$$

or

$$Y(z) = \frac{b_0}{1 - \sum_{1}^{N} a_i z^{-i}} X(z) \tag{5}$$

Thus, the autoregressive data model corresponds to an all-pole system model.

The moving average data model is given by

$$y_n = b_0 x_n + b_1 x_{n-1} + \ldots + b_k x_{n-k}$$

$$= \sum_{0}^{k} b_i x_{n-i} \tag{6}$$

or

$$Y(z) = \left[\sum_{0}^{k} b_i z^{-i} \right] X(z) \tag{7}$$

which says the moving average data model is an all-zero system model.

Finally, the mixed-data model is the autoregressive-moving average data model given by

$$y_n = \sum_{1}^{N} a_i y_{n-i} + \sum_{0}^{k} b_i x_{n-i} \tag{8}$$

or

$$Y(z) = \frac{\displaystyle\sum_{0}^{k} b_i z^{-i}}{1 - \displaystyle\sum_{1}^{N} a_i z^{-i}} X(z) \qquad (9)$$

which is a pole-zero system model.

A special case, which we shall discuss at length later, is to attempt to predict y_n using a linear combination of past values of y_n, e.g.,

$$\hat{y}_n = \sum_{1}^{N} a_i y_{n-i} \qquad (10)$$

There will, of course, be an error between this predicted value and the actual value, i.e.,

$$e_n = y_n - \hat{y}_n = y_n - \sum_{1}^{N} a_i y_{n-i} \qquad (11a)$$

or

$$y_n = \hat{y}_n + e_n \qquad (11b)$$

which may be expressed in the z domain as

$$Y(z) = \frac{1}{1 - \displaystyle\sum_{1}^{N} a_i z^{-i}} E(z) \qquad (12)$$

This is an AR model, where e_n may be considered the input or driving function as depicted in Figure 1A. Alternately, in Figure 1B y_n is the input and $\hat{y}_n$ is the output. In the latter case, we attempt to select the a_i in order to minimize the error, e_n, at least for the linear prediction application.

In either case, the a_i are usually found by minimizing the mean-square error. These coefficients are known as the autoregressive coefficients; they are the same as the linear predictive (LP) coefficients and are also known as the maximum entropy (ME) coefficients. But we shall discuss this at length in the following sections.

III. SPECTRAL ANALYSIS — PREVIOUS PROCEDURES[1]

For discrete time series data there are two equivalent definitions for the power spectrum. If we denote

$$X_N(z) = \sum_{0}^{N-1} x_n z^{-n} \qquad (13)$$

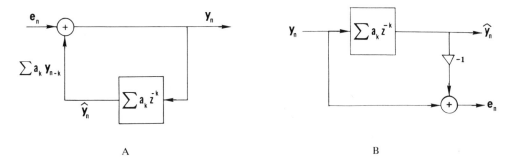

FIGURE 1. Block diagram of error equation; (A) error function as excitation, model commonly used for speech; (B) output prediction based on past input values; coefficients are determined by minimizing the error, e_n.

then the power spectrum is

$$S(\omega) = \lim_{N \to \infty} \frac{1}{N} E|X_N(z)|^2, \quad z = e^{j\omega t} \tag{14}$$

where E denotes the expected value. This is the *direct* definition of the power spectrum, since it is based on the z-transform of the data evaluated on the unit circle. This is rather easily shown to be equivalent to the z-transform of the autocorrelation function

$$S(\omega) = \lim_{N \to \infty} \sum_{-(N-1)}^{N-1} (1 - \frac{|m|}{N}) R_m z^{-m} \tag{15}$$

where the autocorrelation function is defined as

$$R_m = R(mT) = E(x_n x_{n-m}) \tag{16}$$

This is known as the autocorrelation definition, since it is based on the z-transform of the autocorrelation function evaluated on the unit circle.

In practice, one can never perform *exact* ensemble averages of the data, rather one must resort to calculating estimates using a single member function from the ensemble. The expected value of the data is replaced with an estimation procedure such as time averaging. One commonly used estimate is the periodogram.

$$\hat{S}(\omega) = \frac{1}{N} |x_N(\omega)|^2 \tag{17}$$

The quality of the spectral *estimate* is then evaluated by determining whether the estimate is biased or not and by determining if the variance of the estimate becomes smaller as the data record length (number of data samples, N, for a fixed sampling interval, T) increases. The bias is defined as the difference between the expected value of the estimate and the true value of the quantity being estimated, which in this case is the true power spectrum.

Two commonly used estimates for the autocorrelation function are

$$\hat{R}_k = \frac{1}{N-|k|} \sum_{0}^{N-|k|-1} x_n x_{n+|k|} \tag{18}$$

which is unbiased. And

$$\hat{R}_k = \frac{1}{N} \sum_{0}^{N - |k| - 1} x_n \, x_{n+|k|} \qquad (19)$$

which is biased, since

$$E[\hat{R}_k] = \frac{N - |k|}{N} R_k \qquad (19a)$$

The variance of these estimates is discussed in most modern textbooks on signal processing. The estimate selected for the autocorrelation function is usually the one with the minimum bias and minimum variance relative to the true autocorrelation function. Such estimates are said to have good statistical stability and yield good power spectral estimates.

The statistical stability of the spectral estimate may also be increased by windowing the estimate of the autocorrelation function before transforming to the frequency domain. Windowing will generally reduce the variance of the power spectral estimate. But the various window functions are generally unrelated to the data or the random process being analyzed.

Two factors affecting the spectral resolution of the spectral estimate are the finite record length of the autocorrelation function (and the data) and the windowing process applied to the autocorrelation function. As the record length increases, the spectral resolution usually increases (but not always, e.g., the periodogram of white noise just becomes oscillatory). And if the window is designed to reduce the leakage (or sidelobes) in the spectral domain, then the spectral resolution is decreased.

Until recently, increased spectral resolution was achieved by appending a sequence of zeros to the windowed autocorrelation function estimate prior to transforming to the frequency domain. But this only decreases the spacing between spectral line components in the discrete Fourier transform and does not really improve the resolution between two closely spaced spectral components of the signal. The more modern techniques of autoregressive (AR), linear predictive (LP), and maximum entropy (ME) spectral analysis can achieve increased spectral resolution, particularly for short data records. These equivalent methods in effect extend the autocorrelation function by extrapolation or prediction rather than appending zeros.

Besides windowing, the variance of the power spectral estimate may be reduced by segmental averaging, which segments the data record and calculates an estimate of the autocorrelation function for each data segment. The average of these autocorrelation estimates is then determined. The estimated power spectrum is the discrete Fourier transform (DFT) of the average autocorrelation estimates. These power spectral estimates are biased, and the bias exceeds the bias of the periodogram of the unsegmented data record. This is because the main lobe of the segmented spectral window is broader than that of the window used for the periodogram. However, the variance of the spectral estimate of the segmented data is less than the variance of the periodogram by a factor equal to the number of segments, if the segments can be considered independent (or uncorrelated, Gaussian segments). But the segmented estimate has less resolution than that of the periodogram of the entire data record.

Bartlett's procedure[2,3] for power spectral estimation is segmental averaging, but it is performed in the frequency domain. This technique calculates the periodogram of individual segments of the data record and then averages the individual periodograms.

Welch's method[1,3] segments the data, windows each segment, calculates the periodogram of each windowed segment, and then determines the averaged periodogram. The resultant power spectral estimate is called the modified periodogram. The data segments may be juxtaposed or overlapped. The variance of this spectral estimate is less than that of the periodogram, but the resolution is reduced.

We summarize the various procedures for estimating the power spectrum below. The two major categories are the autocorrelation and the direct methods.

IV. AUTOCORRELATION METHODS

1. Estimate the autocorrelation function and then transform

$$\hat{S}(\omega) = \sum_0^L \hat{R}_n e^{-j\omega n} \tag{20}$$

where L is the maximum value of the autocorrelation lags and is usually restricted to 10% of the available number of data samples, i.e., $L \simeq N/10$. This is known as the *Blackman-Tukey* method and is easily seen to be an all-zero or MA model. A variation on this method is to window the data prior to calculating $\hat{R}_n$.

2. Estimate the autocorrelation function of the unwindowed data, window, and transform

$$\hat{S}(\omega) = \sum_0^L \hat{R}_n w_n e^{-j\omega n} \tag{21}$$

This procedure may yield a negative power spectrum estimate unless the transform of the window is positive for all frequencies. This is the generalized Blackman-Tukey method.

3. Segment the data record into K sections, each M points long, such that $N = KM$. Estimate the autocorrelation function for each segment. Average the individual autocorrelation estimates and then transform.

4. Repeat Step 3, but apply a window to each segmental autocorrelation segment, then average and transform.

Zeros may be appended to the autocorrelation estimates prior to transforming (but after windowing) to increase "resolution" in the frequency domain.

V. DIRECT METHODS

The direct methods may be summarized as follows:

1. Calculate the periodogram
   a. Using the rectangular data window
   b. Using another data window.

$$\hat{S}(\omega) = \frac{1}{U} \frac{1}{N} \left| \sum_0^{N-1} x_n w_n e^{-j\omega n} \right|^2 \tag{22}$$

where the window function is such that

$$U = \frac{1}{N} \sum_{0}^{N-1} w_n^2 \qquad (22a)$$

U is usually selected as unity.

2. Smooth the periodogram
    a. Average adjacent periodogram values using equal weighting
    b. Repeat Step a using unequal weighting (a form of moving-average smoothing)
3. Segment the data record into K segments, each M samples long. Calculate the periodogram of each segment and average. The segments may be juxtaposed (N = KM) or overlapped (K>N/M). Overlapping will reduce the variance of the estimate.
4. Segment the data record, window each segment, calculate the periodogram of each windowed segment and average. The segments may be juxtaposed or overlapped. Overlapping reduces the variance of the estimate but the segments are less independent and more correlated. This is Welch's method, which is the "best" of the direct methods and is recommended over the autocorrelation methods. Here

$$\hat{S}_i(\omega) = \frac{1}{UM} \left| \sum_{0}^{M-1} x_{n+iM} w_n e^{-j\omega n} \right|^2 \qquad (23)$$

where $o \leqslant i \leqslant K-1$, $K = N/M$. When the data segments are overlapped, replace M with CM in x(n + iM) where $o \leqslant C \leqslant 1$. The window should be normalized as in Equation 22.

The most commonly used windows are the following:

1. Rectangular window

$$w_n = 1, \ 0 \leqslant n \leqslant N-1 \qquad (24)$$

2. Hanning window

$$w_n = \frac{1}{2} \left(1 - \cos \frac{2\Pi n}{N} \right), \ 0 \leqslant n \leqslant N-1 \qquad (25)$$

3. Hamming window

$$w_n = 0.54 - 0.46 \cos \frac{2\Pi n}{N}, \ 0 \leqslant n \leqslant N-1 \qquad (26)$$

4. Bartlett (or triangular) window

$$w_n = \left[1 - 2 \frac{|n - N/2|}{N} \right], 0 \leqslant n \leqslant N-1 \qquad (27)$$

and n even.

5.   Tukey (or Taper) window

   The leading and trailing 10% ends of the data are multiplied by a raised cosine, i.e., the leading edge by 1/2 (1-cos $2\pi n/M$) and the trailing edge by the mirror image, where $M = N/10$.

One or more of these procedures is presently used by the majority of investigators. But the more modern methods, introduced about 10 years ago, are being used increasingly.

Several of these methods* are compared in Figure 2, for simulated data consisting of the sum of two sinusoids plus random noise. Here we show the autocorrelation function of the sum of the two sinusoids and the spectra calculated using the Blackman-Tukey method and all of the five windows described above, with the exception of the Hamming window. These spectra are compared with the periodogram for several signal-to-noise ratios. As the signal-to-noise ratio decreases, the "best" result is achieved using the Hanning window. The other spectral estimates do not give adequate resolution of the two sinusoidal spectral peaks. All autocorrelation estimates were calculated using Equation 18 to 64 lags with 1024 input data points. The spectra are shown from zero Hz to one-half the sampling frequency, or 64 data points.

VI. MODERN METHODS

About 10 years ago, two nonlinear spectral estimation procedures were introduced, the AR, LP, ME equivalent methods and the maximum likelihood method (ML or MLM).[1] From the point of view of spectral analysis, these methods are particularly attractive for making high resolution spectral estimates when the data record is short. Neither method uses window functions, and the methods adjust themselves to be least disturbed by power at other frequencies.[1] Thus, they are considered data adaptive.

The ME method predicts the autocorrelation function beyond the data limited range. The principle used for this prediction process is that the spectral estimate must be the most random or have the maximum entropy of any power spectrum that is consistent with the known sample values of the autocorrelation function. The objective is to add no new information.

This procedure has been shown to be equivalent to least squares error linear prediction and autoregression.

The model that most easily illustrates the derivation of the appropriate equations is that of linear prediction. Here the predicted value of the nth sample is

$$\hat{x}_n = \sum_{1}^{p} a_i x_{n-i} \tag{28}$$

and the corresponding error is

$$e_n = x_n - \hat{x}_n = x_n - \sum_{1}^{p} a_i x_{n-i} \tag{29}$$

---

*   Many of the methods used were provided by R. Menendez, now employed by Harris Corp., Melbourne, Fla.

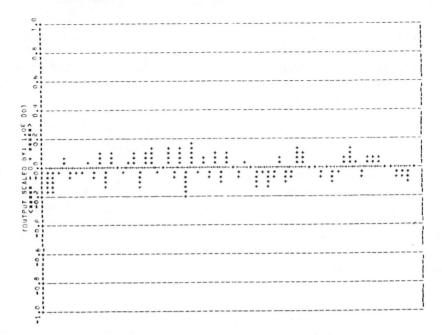

FIGURE 2A. Data sequence, S/N = 12.5 dB.

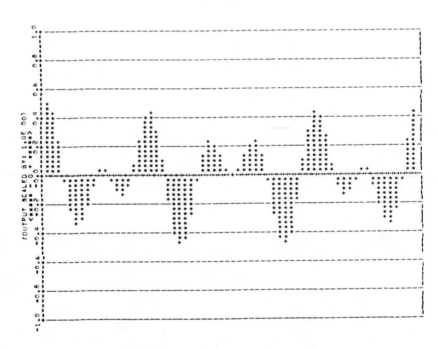

FIGURE 2B. Signal autocorrelation, no noise.

FIGURES 2A-2T. Data, autocorrelation functions, and spectra for two sinusoids plus noise. The two sinusoids are located at $9f_s/128$ and $15f_s/128$. The point farthest to the right on the frequency axis is $f_s/2$ (64 points).

The autoregressive model is

$$x_n = \hat{x}_n + e_n \qquad (30)$$

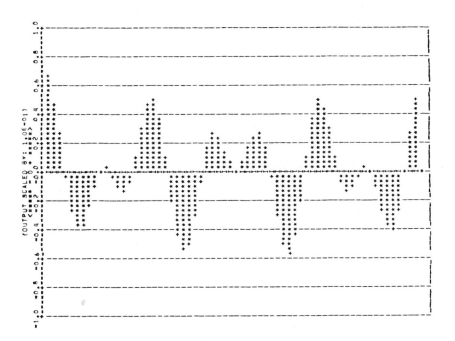

FIGURE 2C.   Signal plus noise autocorrelation, S/N = 12.5 dB.

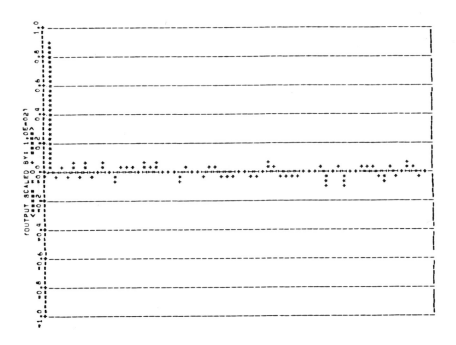

FIGURE 2D.   Noise autocorrelation.

where the error term is orthogonal in an expected value sense to all past data samples. We shall soon show this.

The actual values for the a_i in this parametric model are those found by using expec-

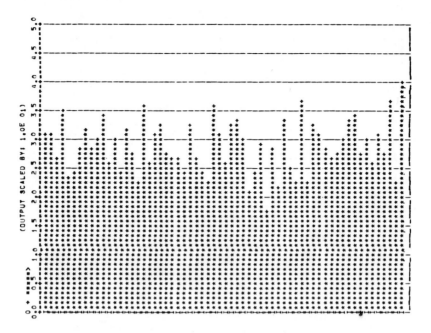

FIGURE 2E.   Periodogram of noise.

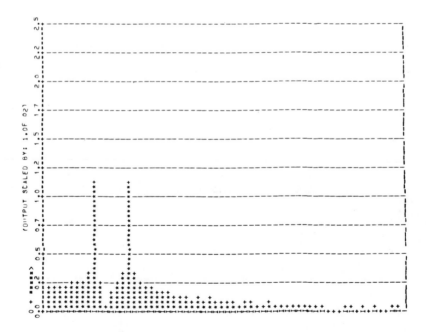

FIGURE 2F.   Periodogram of signal, no noise.

tation, i.e., we minimize the mean-square error

$$E[e_n^2] = E[x_n - \hat{x}_n]^2 \qquad (31)$$

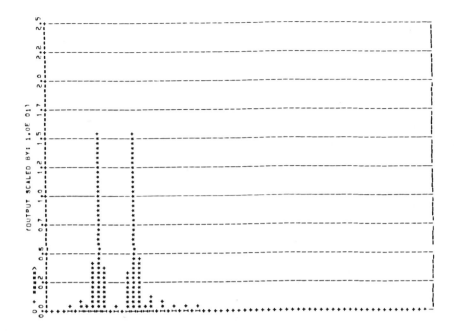

FIGURE 2G.   Blackman-Tukey spectrum, rectangular window, no noise.

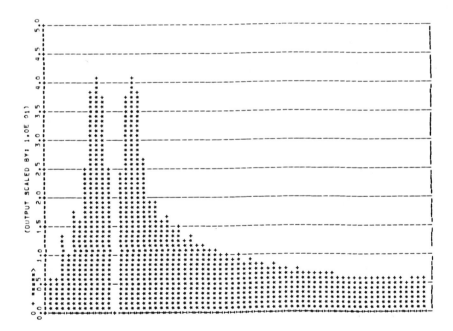

.FIGURE 2H.   Blackman-Tukey spectrum, Hanning window, no noise.

This is achieved by

$$\frac{\partial E[e_n^2]}{\partial a_k} = 0, k = 1, \ldots, p \tag{32}$$

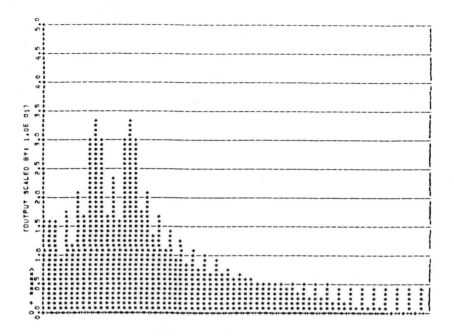

FIGURE 2I.   Blackman-Tukey spectrum, triangular window, no noise.

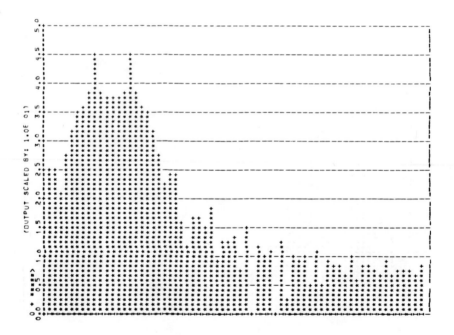

FIGURE 2J.   Blackman-Tukey spectrum, Tukey window, no noise.

which gives

$$E[(x_n - \sum_{1}^{p} a_i x_{n-i})(x_{n-k})] = 0 \qquad (33)$$

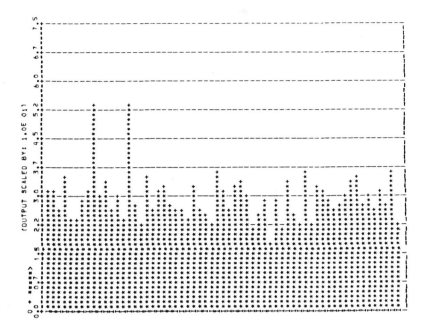

FIGURE 2K.   Periodogram, S/N = 12.5 dB.

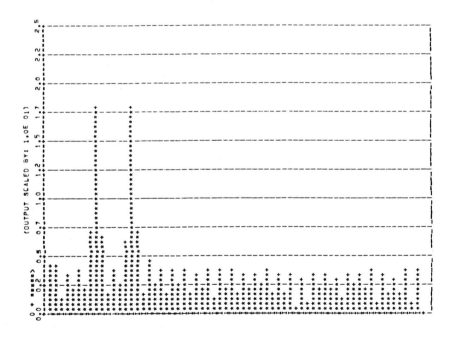

FIGURE 2L.   Blackman-Tukey spectrum, rectangular window, S/N = 12.5 dB.

or

$$R_k - \sum_1^p a_i R_{|k-i|} = 0, \ k = 1, \ldots, p \tag{34}$$

One additional equation is the mean-square error, i.e.,

$$E[e_n{}^2] \ = \ \sigma_e{}^2 \ = \ E[(x_n - \sum_1^p a_i x_{n-i})^2]$$

$$= \ E[(x_n - \sum_1^p a_i x_{n-i})x_n - (x_n - \sum_1^p a_i x_{n-i}) \sum_1^p a_i x_{n-i}]$$

$$= \ E[(x_n - \sum_1^p a_i x_{n-i})x_n]$$

$$= \ R_0 - \sum_1^p a_i R_{|i|} \tag{35}$$

Equations 34 and 35 are known as the Yule-Walker equations.[1]

These $(p+1)$ equations determine the (p) a_i and the mean-square error, $E(e_n{}^2)$. In matrix form we have

$$[R] \ \vec{\lambda} \ = \ \vec{\sigma^2} \tag{36}$$

where the autocorrelation matrix is

$$[R] \ = \ \begin{bmatrix} R_0 \ R_1 \ R_2 \ \ldots \ R_p \\ R_1 \ R_0 \ R_1 \ \ldots \ R_{p-1} \\ \vdots \\ R_p \cdot \ldots \ldots \ldots R_0 \end{bmatrix} \tag{37}$$

and $\vec{\lambda}$ is the column vector $[1, -a_1, -a_2, \ldots, -a_p]$ while $\vec{\sigma^2}$ is the column vector $[\sigma^2{}_e, 0, 0, \ldots, 0]$.

Thus

$$\vec{\lambda} \ = \ [R]^{-1} \ \vec{\sigma^2} \tag{38}$$

The orthogonality principle[4] has been used in this derivation, namely that

$$E[e_n x_{n-i}] \ = \ 0, i \ = \ 1, \ldots p$$

We may derive this principle by interchanging the order of expectation and differentiation. We have from Equation 32 that

$$\frac{\partial E(e_n{}^2)}{\partial a_k} \ = \ 0, k=1, \ldots, p$$

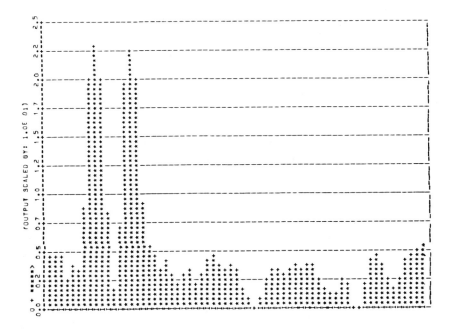

FIGURE 2M. Blackman-Tukey spectrum, Hanning window, S/N = 12.5 dB.

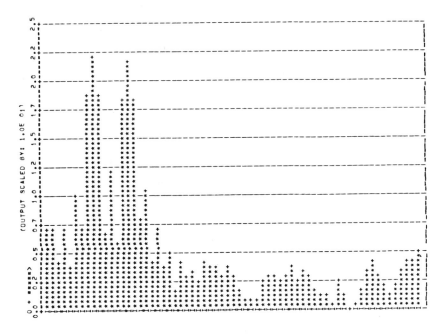

FIGURE 2N. Blackman-Tukey spectrum, trinagular window, S/N = 12.5 dB.

But this may be written as

$$E \frac{[\partial e_n{}^2]}{\partial a_k} = 2 \ E\left[e_n \ \frac{\partial e_n}{\partial a_k}\right] = 2E \ [e_n x_{n-k}(-1)]$$

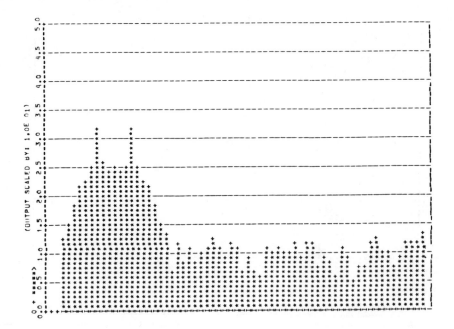

FIGURE 2O.   Blackman-Tukey spectrum, Tukey window, S/N = 12.5 dB.

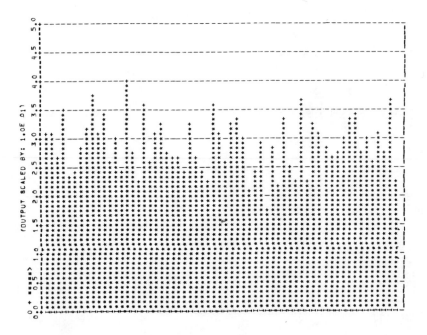

FIGURE 2P.   Periodogram, S/N = −3 dB.

Thus:

$$E\,[e_n x_{n-k}] \ = \ 0, k=1,\ldots,p \qquad\qquad (39)$$

This says that the expected value of the product of the error ($e_n \ = \ x_n - \hat{x}_n$) with each

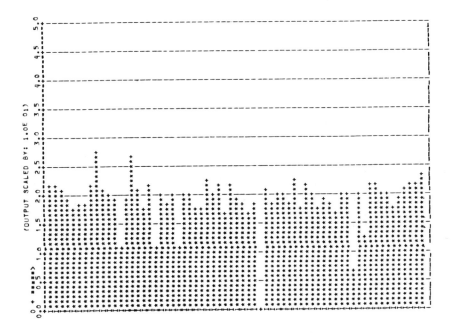

FIGURE 2Q. Blackman-Tukey spectrum, rectangular window, S/N = −3 dB.

known data sample x_{n-i} is zero, i.e., the error is orthogonal to the known data samples in an expected value sense, i.e., geometrically $e_n + \hat{x}_n = x_n$.

The z-transform may be obtained by transforming Equation 30, using Equation 28, i.e.,

$$X(z) = \frac{E(z)}{1 - \sum_{1}^{p} a_i z^{-i}}$$

$$= \frac{E(z)}{\sum_{0}^{p} \lambda_i z^{-i}} = \frac{E(z)}{\vec{\lambda}^T \vec{Z}}$$

where

$$\vec{\lambda} = \text{col} \, [1, -a_i, -a_2, \ldots, -a_p]$$

$$\vec{Z} = \text{col} \, [1, z^{-1}, z^{-2}, \ldots, z^{-p}] \tag{40}$$

and the superscript T denotes transpose.

The power spectrum is given as

$$\lim_{N \to \infty} \frac{1}{N} E|X(z)|^2 = \lim_{N \to \infty} \frac{1}{N} \frac{E|E(z)|^2}{\vec{Z}^{*T} \vec{\lambda}^* \, \vec{\lambda}^T \vec{Z}} = \lim_{N \to \infty} \frac{1}{N} \frac{E|E(z)|^2}{|\vec{\lambda}^T \vec{Z}|^2}$$

$$= \lim_{N \to \infty} \frac{1}{N} \frac{E|E(z)|^2}{\left| 1 - \sum_{1}^{p} a_i a^{-i} \right|^2} \tag{41}$$

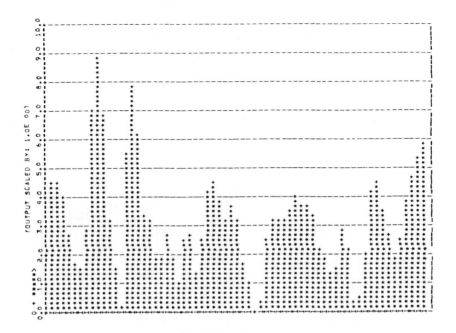

FIGURE 2R.   Blackman-Tukey spectrum, Hanning window, S/N = −3 dB.

where the * denotes complex conjugate.

The numberator of Equation 41 may be found by recalling that since e_n is a white noise (innovative) process, then

$$\lim_{N \to \infty} \frac{1}{N} E|E(z)|^2 \ (2B) = \sigma_e^2 \tag{42}$$

where B is the one-sided spectral bandwidth of the process such that the sampling interval, T = 1/2B. Then

$$\lim_{N \to \infty} \frac{1}{N} E|X(z)|^2 = \frac{\sigma_e^2 T}{\left| 1 - \sum_1^p a_i z^{-i} \right|^2} \tag{43}$$

In practice, we typically *estimate* the spectrum by simply using $|X(z)|^2$, i.e., dropping the limiting operation and the expectation and replacing the a_i with their estimates $\hat{a}_i$.

We point out that Equation 40 may be interpreted as

$$E(z) H(z) = X(z) \tag{44}$$

i.e., the white noise process is exciting a filter H(z) to produce an output X(z). Such a model is used in speech research where

$$H(z) = \frac{1}{1 - \sum_1^p a_i z^{-i}} \tag{45}$$

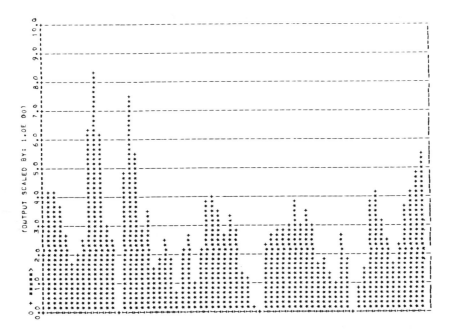

FIGURE 2S.　Blackman-Tukey spectrum, triangular window, S/N = −3 dB.

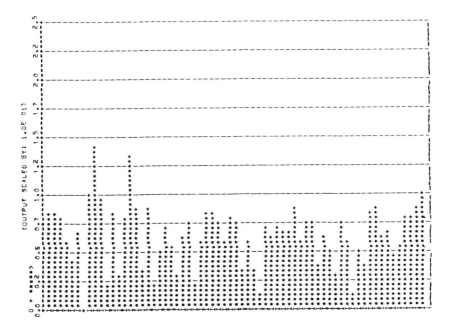

FIGURE 2T.　Blackman-Tukey spectrum, Tukey window, S/N = −3 dB.

is called the *forward filter* or vocal tract filter model and X(z) is the transform of the speech signal. E(z) then represents the transform of the glottal volume-velocity of air, expelled from the lungs through the vocal cords to excite the vocal tract.

The *inverse filter* is the reciprocal of H(z). And the inverse filtering problem is to estimate the glottal volume-velocity signal by inverse filtering the monitored speech signal. We shall illustrate this later.

In practice we must estimate the a_i, and we denote these estimates as $\hat{a}_i$. It is common practice to minimize the total squared error in order that the estimation equations will be similar to the statistical expectation equations. Further, we may generalize our prediction equation to consider data consisting of signal plus noise. Our linear prediction filter will then also estimate the signal as well. This is the discrete counterpart of Wiener's linear prediction-estimation problem.

We now have $x_n = s_n + n_n$ and that

$$\hat{s}_n = \sum_1^P a_i x_{n-i}.$$

We define the total squared error as

$$E = \sum_{n=0}^{N-1} e_n^2 = \sum_0^{N-1} (s_n - \hat{s}_n)^2$$

$$= \sum_0^{N-1} \left(s_n - \sum_1^p a_i x_{n-i} \right)^2 \tag{46}$$

Note that if our data were noise free, then $x_n = s_n$ and we would have just the prediction problem. For this case we have

$$E = \sum_0^{N-1} \left(s_n - \sum_1^p a_i s_{n-i} \right)^2$$

$$= \sum_0^{N-1} \left(x_n - \sum_1^p a_i x_{n-i} \right)^2 \tag{47}$$

We determine the estimates for a_i by setting the partial derivatives of E with respect to a_k to zero,

$$\frac{\partial E}{\partial a_k} = 0, 1 \leqslant k \leqslant p \tag{48}$$

This gives

$$\sum_{i=1}^p \hat{a}_i \sum_{n=0}^{N-1} x_{n-i} x_{n-k} = \sum_0^{N-1} s_n x_{n-k} \tag{49}$$

for $1 \leqslant k \leqslant p$. The $\hat{a}_i$ are found by solving Equation 49.

Note that the inner summation on the left is an estimate for the autocorrelation function of the noisy data and that the summation on the right is an estimate of the cross-correlation function of the signal with the noisy data.

Several cases may be considered. First, we may assume that in fact we have $p + N$ consecutive samples available for computation, but we only calculate our autocorrelation and cross-correlation estimates using N samples so that we may avoid end effects. This will give us p linear equations with p unknown coefficients $\hat{a}_i$. This would give

$$\sum_1^p \hat{a}_i \hat{R}_{xx} (i-k) = \hat{R}_{sx} (k), 1 \leq k \leq p \qquad (50)$$

where

$$\hat{R}_{xx} (i-k) = \sum_0^{N-1} x_{n-i} x_{n-k} \qquad (51)$$

$$\hat{R}_{sx} (k) = \sum_0^{N-1} s_n x_{n-k} \qquad (52)$$

We could divide both Equations 51 and 52 on the right by N if desired.

Second, we may window the data sequence with an N point long window function and assume we have only N consecutive samples available for computation. Then we have

$$\sum_{i=1}^p \hat{a}_i \hat{R}_{xx} (i-k) = \hat{R}_{sx} (k), 1 \leq k \leq p \qquad (53)$$

where

$$\hat{R}_{xx} (i-k) = \sum_0^{N-|k|-1} (x_{n-i} w_{n-i}) (x_{n-k} w_{n-k}) \qquad (54)$$

$$\hat{R}_{sx} (k) = \sum_0^{N-|k|-1} s_n (x_{n-k} w_{n-k}) \qquad (55)$$

and w_n is the window function selected. We can, of course, divide Equations 51 and 52 on the right by the constant $N-|k|$ if desired. Further, the desired signal, s_n, may also be windowed as well. But in this formulation it seems inappropriate to do so, while in speech processing s_n is usually considered synonymous with the data and is thus replaced with $s_n w_n$.

A third technique for calculating $\hat{a}_i$ is known as Burg's method,[1] which is commonly used in processing seismic data. Burg's method formulates an average mean square error determined by averaging the error found by processing the data in the forward direction with the prediction filter with that error calculated by processing the data in the reverse direction. The coefficients determined by Burg's method differ slightly from those found by the other procedure. Further, there is no need to estimate the autocorrelation function.

Equation 50 or 53 represent the discrete Wiener filter equations and we may solve for the discrete Wiener filter using transform methods. However, one normally proceeds as follows.

We may express either Equation 50 or Equation 53 in matrix notation as

$$[\hat{R}_{xx}] \ \vec{\hat{a}} = \vec{\hat{R}}_{sx} \tag{56}$$

so that

$$\vec{\hat{a}} = [\hat{R}_{xx}]^{-1} \ \vec{\hat{R}}_{sx} \tag{57}$$

In neither the estimation nor the prediction problem do we know the true value of the signal, s_n. In speech processing, s_n is replaced with the next data sample x_n, and the a_i coefficients are calculated using the past data samples. Typically, these coefficients are then used to reduce the speech data, i.e., the coefficients are transmitted rather than the sampled speech waveform. At the receiving end the coefficients are used to resynthesize the original speech. This process may achieve a data reduction by a factor of 20, since the speech signal is approximately stationary for 25 msec and approximately 10 to 15 coefficients may be used to represent a 25 msec signal sampled at, say, 10 KHz.

The estimated autocorrelation matrix $[\hat{R}_{xx}]$ is a particular type known as a Toeplitz matrix, which is one with equal elements along any diagonal; further, it is symmetric and positive definite. Special algorithms exist for solving for $\vec{a}$ by recursive means so that $[\hat{R}_{xx}]$ does not have to be inverted.[1]

The remaining difficulties in the calculation of the $\hat{a}_i$ are selecting an estimator for the elements of the $\hat{R}_{xx}$ matrix and the $\hat{R}_{sx}$ vector. Several autocorrelation estimators have been described earlier in the chapter. One cross-correlation estimator, as mentioned above, is to replace the true or desired signal, s_n, with the data x_n (or its windowed counterpart). Another cross-correlation estimator is replace s_n and x_n with an average.[5] This is possible if the experiment may be repeated a number of times as in some evoked response studies. In this case we have

$$\hat{R}_{sx}(k) = \sum_{0}^{N-1} \tilde{s}_n \tilde{s}_{n-k}, \ 1 \leqslant k \leqslant p \tag{58}$$

where

$$\tilde{s}_n = \tilde{s}(n) = \frac{1}{M} \sum_{j=1}^{M} x_j(n) \tag{59}$$

where $x_j(n)$ denotes the jth repetition of the data at time n and there are M such repetitions. The signal in the data repetitions is assumed to be time locked (or approximately so) from repetition to repetition.

No matter which method is selected for estimating the $\hat{a}_i$, once they are determined they may be used to:

1.   Predict the data and/or estimate the signal

2.   Predict the autocorrelation function (as can be seen from Equations 50 and 53)
3.   Estimate the power spectrum

In the latter case, we estimate the power spectrum of the data with

$$|X_{(z)}|^2 = \frac{\sigma_e^2 T}{\left|1 - \sum_1^p \hat{a}_i z^{-i}\right|^2} \tag{60}$$

where the $\hat{a}_i$ are determined by one of the previously discussed estimation techniques.

This spectral estimate is known as either the linear predictive, autoregressive, or maximum entropy estimate.[1] Note that the reciprocal of the magnitude squared of the transform of the linear predictive coefficients is the power spectrum. The spectrum is determined by the coefficients, and in practice it is not necessary to actually predict the autocorrelation values and then transform the sequence of calculated and predicted autocorrelation values. This is because the data model says that the data may be represented by the coefficients.

The maximum likelihood (ML) spectral estimate is related to the ME estimate. The ML estimate is a minimum variance unbiased estimator of the spectrum and may be derived by solving a classical optimum filtering problem.[1] The inverse transform of the ML estimate does not in general agree with the measured autocorrelation values. The ML estimate is in fact an average of various ME estimates, i.e., if the ME spectra are estimated for $k = 1,...,p$ and the average of the reciprocals of the spectra are determined, then this average is equal to the reciprocal of the ML spectrum.[1] This implies that the ML spectral estimate is more stable statistically but has less spectral resolution than the ME spectral estimate.

A major problem with these autoregressive data models is selecting or determining the "proper" value of p, i.e., the number of filter coefficients. This is discussed at length elsewhere.[1] If p is too small, then the resulting spectral estimate is highly smooth; while if p is too large, then spurious detail is introduced in the spectrum. This is illustrated in Figure 3. One criterion for determining p is to calculate the mean-square error as p is increased. When the error suddenly decreases by a significant step, then select this value of p. Other criteria are available.[1]

To date no good estimate of the variance of the ME estimator exists. More research is needed in this area.

The advantage of the LP/AR/ME approach is that a formal all-pole model is applied as an estimate of the spectrum. The resolution of this estimator is greater than previous methods, but the estimator is highly sensitive to the signal-to-noise ratio of the data. Further, the computational complexity of the method is on the same order as that for the older more established methods. Finally, the LP method is data adaptive because of the model, i.e., new coefficients are determined for each data set. No windows are needed either.

These new techniques are finding wide application, including linear arrays of sensors and image processing.[1]

As a brief summary we compare some properties of the LP/AR/ME spectral estimator with those of other methods.[1,9]

Estimation of autocorrelation function

ME   Uses known or estimated autocorrelation values unmodified. Predicts unknown values.

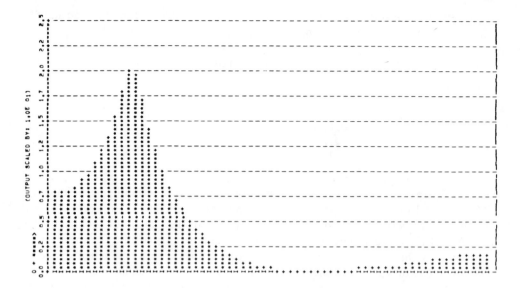

FIGURE 3A.   p = 5, S/N = 12.5 dB.

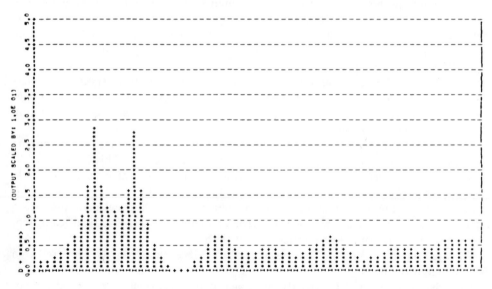

FIGURE 3B.   p = 15, S/N = 12.5 dB.

FIGURES 3A-E. Linear predictive, autoregressive spectra. The two sinusoids are located at $9f_s/128$ and $15f_s/128$. The point farthest to the right on the frequency axis is $f_s/2$ (64 points).

DFT   Usually windows or weights autocorrelation values before transforming. Introduces leakage in the spectrum.

Spectral window effects

ME   No window effects are introduced since the autocorrelation values are not windowed. However, if the AR model does not fit the data well, then similar effects can occur, i.e., error is not white noise.

DFT   Window effects always present.

Estimates of peak power of a sinewave of power P

ME   Proportional to $P^2 L^2$, where L is the number of autocorrelation values used or estimated.

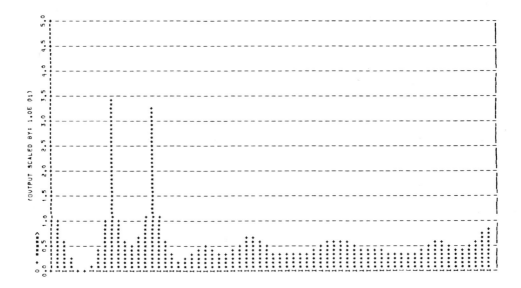

FIGURE 3C.   p = 20, S/N = 12.5 dB.

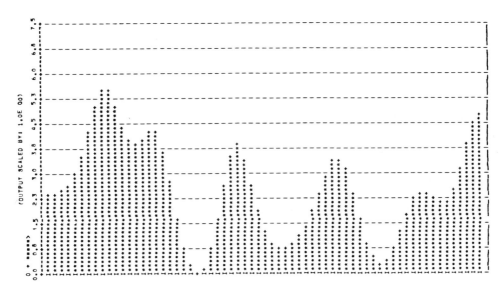

FIGURE 3D.   p = 15, S/N = −3 dB.

DFT	Proportional to P
ML	Proportional to P

Estimate of bandwidth of spectral peak of sinewave (W)

ME	Proportional to $1/PL^2$
DFT	Proportional to $1/L$
ML	Proportional to $1/\sqrt{p}$ · $1/L\sqrt{L}$

Estimate of spectral power in sinewave (W P)

ME	Proportional to P
DFT	Proportional to P/L
ML	Proportional to $\sqrt{P}$ / $L\sqrt{L}$

Spectral reliability

ME   More work needed. Asymptotically, the variance approaches that of the

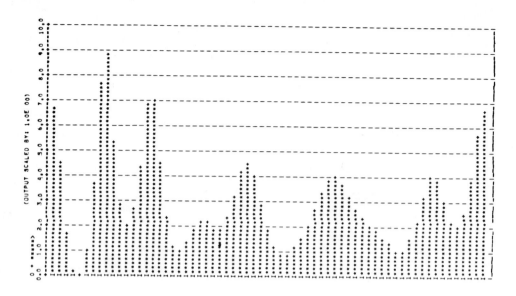

FIGURE 3E.   p = 20, S/N = −3 dB.

other methods. The degrees of freedom, K, (twice the ratio of the square of the mean of the spectral estimate to the variance of the spectral estimate) is less than N/L where N is the number of data samples.

DFT $K \cong 2N/L$ (Bartlett window)

          $K \cong N/L$ (rectangular window)

ML $K \cong N/L$

Linearity of spectral estimate

ME   Estimation is nonlinear, i.e., spectrum of the sum of two time series is not the sum of the spectra of each time series.

DFT   Estimation is linear.

Estimation of separation between spectral lines

ME   Not easily defined. Can be estimated.

DFT   1/LT where T is the sampling interval.

The above results are dependent upon the data and algorithms used.

VII. EXAMPLE — SPEECH

A. Background

The human vocal system is depicted in Figure 4. The lungs are air reservoirs which upon command (for speech) expell air up the trachea to the vocal folds. For voiced sounds (such as vowels), the air pressure increases until the folds are pushed apart forming an opening known as the glottis. A puff of air passes through this opening, setting the vocal folds into vibratory motion. The succession of air pulses generated as a result of this vibratory motion of the vocal cords sets up an acoustic field that continues to travel up the vocal tract. The phonation or sound generated has an auditory correlate (or pitch) that is directly related to the frequency of vibration and loudness that is determined by the amplitude of the acoustic pressure wave. The frequency of oscillation of the folds is determined by their mass, length, thickness, elasticity, and compliance, as well as by the subglottal pressure.

The volume-velocity (rate of air-flow) as it passes through the glottis, modified by

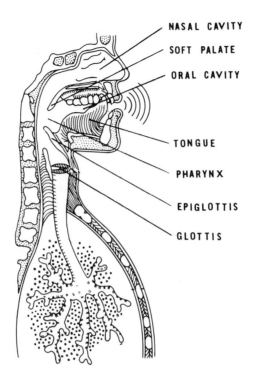

NASAL CAVITY
SOFT PALATE
ORAL CAVITY
TONGUE
PHARYNX
EPIGLOTTIS
GLOTTIS

FIGURE 4.   Anatomy of human vocal tract.

the pharynx, mouth, and nasal cavities, is finally radiated from the lips and nostrils as voiced sounds. All voiced sounds such as vowels and voiced consonants, e.g., b, z, and g, originiate in the vocal cords. The glottal volume-velocity is considered the acoustic source waveform for voiced sounds and has never been directly measured. This waveform is considered a potential indicator of normal or abnormal laryngeal function.[6] For this reason we are interested in estimating the glottal volume-velocity waveform.

Unvoiced sounds occur when the vocal cords are held apart so that the air expelled from the lungs up the trachea is unaffected by glottal vibrators. The two fundamental unvoiced sounds are fricative and plosive. The former is typified by the sound s, which is produced by expelling air through a constriction such as the teeth to produce turbulent air flow; the latter unvoiced sound, typified by p as in puff, is created by the rapid release of air pressure built up behind a closure such as the lips.

The vocal tract is usually considered to be the entire cavity or passageway from the glottis to the lips. We usually assume the source (or glottis) is independent of the vocal tract. The tongue controls the vocal tract dividing it into two resonant cavities, the pharynx and the mouth, which, in turn, determine the transmission characteristics of the vocal tract. The nasal cavity is in parallel with the mouth and can further modify the sound produced.

The transmission characteristic of the vocal tract is usually described by its resonant peaks in the spectrum known as formants, which shift in frequency as we change the characteristics of the vocal tract.

B. Inverse Filtering

If we assume the soft palate or velum is closed so that the nasal passage is closed off, then we may model the human vocal system, as shown in Figure 5, where e_n is the error function used earlier in the chapter.

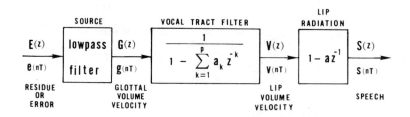

FIGURE 5.   Model of vocal tract.

Using this model, one may synthesize speech by exciting the vocal tract filter with the triangular waveform, shown in Figure 6. For speech synthesis the vocal tract filter and radiation filter coefficients are selected prior to the synthesis process. The speech waveform generated by this synthesis procedure also appears in Figure 6. Now in order to estimate the excitation waveform, in this case the triangular glottal excitation, we assume a linear prediction model, where only the speech signal is assumed known. Using the procedures described earlier in the chapter we derive or estimate the vocal tract linear predictive filter coefficients. In this case, $p = 15$ was required to give a good model.[6] The speech waveform is then inverse filtered using the inverse vocal tract filter to obtain an estimate of the error function, e_n, which is known as the residue function, shown in Figure 6. Finally, the residue or error function is low-pass filtered to give the estimated triangular glottal volume-velocity waveform, shown in Figure 6.[6] We selected this synthesis example to illustrate the method since the answer was known. In practice this is not the case and one must attempt to confirm the estimated waveforms by either processing considerable data or by other means, such as photography.[6]

C. Linear Predictive Spectral Estimation

Another example using speech data is to determine a spectral estimate of the vocal tract, e.g., to attempt to locate the formants for a particular speech utterance. Here the speech record to be analyzed is shown in Figure 7. The signal was sampled at 10 kHz to give 256 data points with a 512 point FFT being calculated. The phonation was a sustained /i/. The data record was Hanning windowed (256 points) prior to transforming. No zeros were appended for this example. The DFT spectrum appears in Figure 7 as the scalloped spectrum. The linear predictive spectrum is shown as the envelope of the DFT spectrum and was calculated from the Hanning windowed data by the autocorrelation method. The windowed data was 20 msec and $p = 16$. The LP spectrum is the reciprocal of the DFT of the LP coefficients. Zeros were appended prior to calculating the 512 point FFT. The scalloping in the standard DFT spectrum is due to the periodically repeated speech waveform introduced by the pitch period.

D. Speech Recognition, Linear Prediction, and Pattern Recognition

As a final example of speech processing we select the problem of speaker recognition; an example provided by Dr. M. Nadal-Suris of the University of Puerto Rico. In this example we shall derive the linear predictive coefficients of vowel sounds for several speakers. If these coefficients cluster, then we may predict the particular vowel phonated by the subject by knowing the location of the cluster of the LP coefficients in the feature space. Further, the clustering may be a function of the speaker. Thus, the location of the LP coefficients may allow a prediction (or estimation) of the actual speaker as well.

For this example, four speakers recorded three vowel sounds all at approximately the same intensity level. The three vowels were

1. /u/ oo as in boot

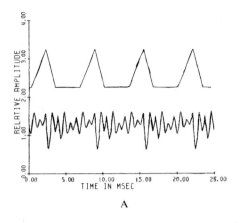 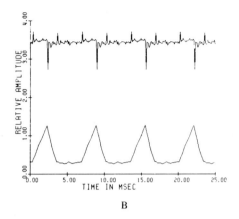

A   B

FIGURE 6.   Example of inverse filtering of synthesized speech; (A) Synthetic glottal-volume velocity wave-form and resultant speech waveform. A 10-pole vocal tract filter was used to generate the speech; (B) Residue or error waveform obtained by inverse filtering and estimated glottal-volume velocity waveform.

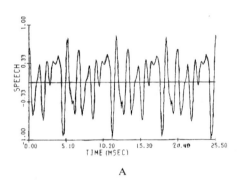 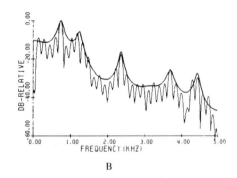

A   B

FIGURE 7.   Example of periodogram and linear predictive spectra for a segment of speech; (A) Speech waveform; (B) periodogram and linear predictive spectra (envelope); the LP spectrum was calculated using the autocorrelation method using 20 msec of Hanning windowed data with p = 16. A 512 point FFT was calculated with zeros appended before transforming.

2. /eI/ strongly stressed a as in late
3. /i/ ee as in seem

We thus have 4×3 = 12 elements in our signal space. The recordings were digitized at 10 KHz and a six-pitch period portion of each signal was analyzed for the LP coeffi-cients with p = 18. Each set of coefficients may be considered an 18 dimensional vector. A cluster analysis of the 12, 18-dimensional vectors was performed to determine if (1) the sounds emitted by each speaker clustered, (2) two different vowel sounds were sufficiently separated in feature space to be distinguishable from one another, and (3) each speaker could be identified from his speech sample.

Several unsuccessful clustering schemes included the calculation of the norm

$$||\vec{\hat{a}}|| = \sum_{i=1}^{n} |\hat{a}_i| \qquad (61)$$

where the LP coefficients vector is $\vec{a} = (\hat{a}_1, \hat{a}_2, \dots \hat{a}_{18})$. This norm was then used to calculate vector distances. The distances for like vowels and different vowels were of the same order of magnitude. An averaging scheme calculated the mean of each pattern and then used this mean as a reference to calculate distances also failed to provide adequate results.

A hyperplane scheme between each pair of pattern classes did prove successful, however. The algorithm used for finding the hyperplanes is known as the perceptron algorithm.[7] This algorithm attempts to linearly separate the pattern classes. If ω_1 and ω_2 are two different pattern classes, which in our case corresponds to two distinct vowel sounds, and w(1) represents an initial weighting vector, arbitrarily chosen, and S_k denote our sounds (S_k, k = 1,2,3,4, the 4 /u/ sounds; k = 5,6,7,8, the 4 /eI/ sounds; and, k = 9,10,11,12, the 4 /i/ sounds), then at the kth training step in the algorithm we have

$$\text{if } S_k \, \epsilon\omega_1 \text{ and } w'(k) \, S_k \leq 0, \text{ replace } w(k) \text{ with}$$

$$w(k+1) = w(k) + cS_k$$

where c is a correction increment and is taken to be positive and $w'(k)$ is the transpose of w(k). If

$$S_k\epsilon\omega_2 \text{ and } w'(k) \, S_k > 0,$$

$$\text{replace } w(k) \text{ with}$$

$$w(k+1) = w(k) - cS_k$$

otherwise, leave w(k) unchanged, i.e.,

$$w(k+1) = w(k)$$

If ω_1 and ω_2 are linearly separable, then the algorithm will converge and the resultant w(k) will be the solution vector.

The algorithm did converge and three hyperplanes were found separating ω_1, and ω_2, ω_1, and ω_3, and ω_2 and ω_3.

While in this example it was possible to linearly separate the pattern classes or sounds for these four speakers, separation may not be possible for a large number of speakers. Speaker recognition for this example was not achieved, since the distance norm did not adequately separate the speakers, possibly because there were too few speakers.

Other pattern recognition techniques could be used, e.g., the autocorrelation coefficients $\hat{R}_i$ could be used in place of the $\hat{a}_i$ or the speech waveform could be approximated by an orthogonal series expansion and a subset of the coefficients of the expansion could be used in place of the LP coefficients.

E. Data Reduction

We draw attention to the fact that another example application of linear prediction is data reduction in speech transmission. A sample calculation of such data reduction was given earlier in the chapter.

VIII. EXAMPLE — EEG

The electroencephalogram (EEG) is the electrical activity of the brain monitored by scalp electrodes.[8] The EEG signal level is typically 10 to 200 μV and occupies a fre-

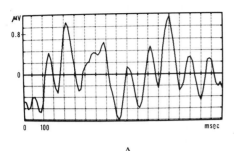

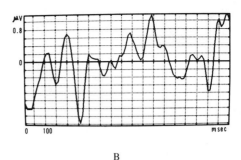

A   B

FIGURE 8. Examples of spectra for EEG, visual evoked response, and an averaged visual evoked response. The sampling frequency was 100 Hz; (A) EEG or background activity, no stimulus; (B) a single VER.

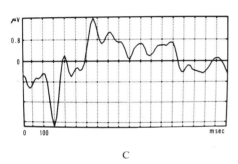

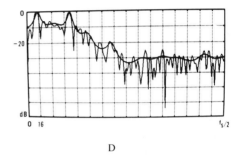

C   D

FIGURE 8C. Average VER for 50 successive stimuli; (D) periodogram and LP spectra (p = 20) of EEG.

quency range from near DC to approximately 30 Hz. The EEG is commonly used in neurological (and other) clinics as a noninvasive diagnostic aid. But brain waves are also studied in sleep research and sensory perception, and cognition laboratories. The evoked response (ER) is a particular type of EEG, namely, the EEG evoked by a particular sensory stimulus, e.g., a flash of light to the eyes, or a tone to the ears. Most EEG or ER research has been concerned with establishing relationships between particular patterns in the brain wave and disease entities to assist the diagnosis of neurological disorders, such as brain damage due to a stroke or tumor or to study brain and sensory modality function by using various stimuli. In brief, investigators have often looked for correlates between the EEG or ER and the state of brain functioning.[8]

A. EEG and ER Spectra

In Figure 8 we show a typical segment of ongoing EEG activity when the subject is not performing a mental task and is not being presented a stimulus. Also shown is a single visual evoked response elicited by a flash of light. The data were sampled at 10 msec intervals and each ER segment is 950 msec long. This figure also shows the average evoked response for 50 repetitions of the stimulus. For each of these signal waveforms we show the periodogram and the LP spectra that we considered to best represent the spectral content.

In Figure 9 we illustrate our spectral analysis methods on a segment of data from a patient with epilepsy. The EEG record shows a quiet, normal segment followed by a spike-and-wave segment typical during a seizure. The periodogram and LP spectra for each of these segments is shown for comparison purposes. The data were recorded by J. Principe at the University of Florida, Gainesville.

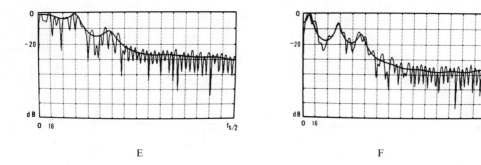

E   F

FIGURE 8E.   Periodogram and LP spectra (p = 15) of single VER; (F) periodogram and LP spectra (p = 15) of average VER.

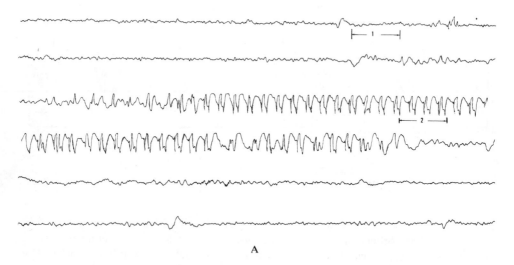

A

FIGURE 9.   EEG and spectra of an epileptic seizure. Sampling rate was 300 Hz; (A) EEG data record showing sections that were spectral analyzed.

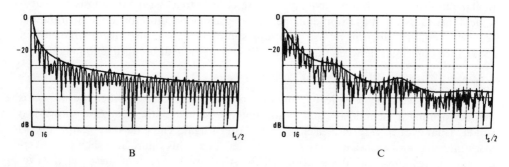

B   C

FIGURE 9B.   Periodogram and LP spectra (p = 5) of Section 1, portion of EEG prior to onset of the seizure; (C) periodogram and LP spectra (p = 10) of Section 2, a portion of the epileptic seizure known as spike-and-wave.

B. EEG Prediction

Our final example illustrates how the EEG may be extrapolated or predicted using the LP coefficients. Figure 10 presents several examples of EEG prediction. Here we

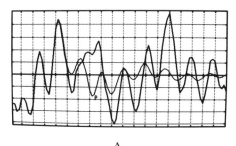
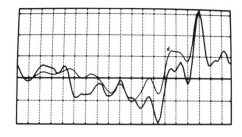

A	B

FIGURE 10. Prediction (extrapolation) of (A) EEG (p = 20) and (B) average visual evoked response (p = 15); the predicted waveform is denoted with a p. The actual waveform is also shown for comparison. The scales are the same as that shown in Figures 8A and 8C.

show the segment of data used to calculate the LP coefficients and the segment of data predicted using these coefficients. The figure also shows the actual data for comparison purposes. We are studying how the predicted EEG, when no stimulus is present, may be subtracted from an ER to improve the signal-to-noise ratio.

IX. SUMMARY

In this chapter we have shown how signal processing has developed recently to model the data with either all poles (AR), all zeros (MA), or a mixture (ARMA). The spectrum was defined and then various estimates of the spectrum were discussed. The spectral representation of the data was then shown to be related to a time domain representation, i.e., the data models have a spectral representation and a time series representation. Various examples with simulated and actual data illustrated how these models could be applied.

Our approach here was to provide an overview of this topic for time series analysis, but considerable work is being done using similar models for array and image processing.[1]

We should point out that the coefficients used in the AR, MA, or ARMA model could be used as features for the data, as is done in speech processing. For example, if such a model characterizes the EEG or evoked response (ER) well, then the coefficients could be used in a vector to classify the EEG or ER waveform for various experimental conditions.

One weakness of these models appears to be that while an AR model may give an excellent representation of the envelope of the spectrum of the data, the coefficients may not actually predict the data well beyond a few points. Thus, the coefficients may need to be more frequently updated for some types of data and less frequently for other types of data.

In fact, one criterion for the adequateness of the model order (the number of coefficients) might be how well the model predicts or extends the data. This is a test easily performed, since the number of coefficients is usually much less than the number of available data samples.

We hope the reader will find the results presented in this chapter a useful reference point for his own data processing, and further, we hope the weaknesses of the results in this area will stimulate research to correct these deficiencies.

REFERENCES

1. **Childers, D. G.**, *Modern Spectrum Analysis*, Institute of Electrical and Electronics Engineers Press, New York, 1978; also distributed by John Wiley & New York.*
2. **Jenkins, G. M. and Watts, D. G.**, *Spectral Analysis and Its Application*, Holden-Day, San Francisco, 1969.
3. **Oppenheim, A. V. and Schafer, R. W.**, *Digital Signal Processing*, Prentice-Hall, Englewood Cliffs, N.J., 1975.
4. **Papoulis, A.**, *Probability, Random Variables, and Stochostic Processes*, McGraw-Hill, New York, 1965.
5. **McGillem, C. D. and Aunon, J. I.**, Measurements of signal components in single visually evoked braon potentials, *IEEE Trans. Biomed. Engr.*, BME-24, 232, 1977.
6. **Childers, D. G.**, Laryngeal pathology detection, *CRC Crit. Rev. Bioeng.*, 2, 373, 1977.
7. **Tou, J. C. and Gonzelez, G. C.**, *Pattern Recognition Principles*, Addison-Wesley, Reading, Mass., 1974.
8. **Childers, D. G.**, Evoked responses: electrogenesis, models, methodology, and wavefront reconstruction and tracking analysis, *Proc. IEEE*, 65, 611, 1977.
9. **Barnard, T. E.**, The Maximum Entropy Spectrum and the Burg Technique, Tech. Rep. No. 1, ALEX(03)-TR-75-01, Texas Instruments, Houston, June 25, 1975.

*   See the Bibliography for extensive references to the literature.

Chapter 4

STATISTICAL PATTERN RECOGNITION

C. H. Chen

TABLE OF CONTENTS

I. INTRODUCTION

In statistical pattern recognition, patterns are considered as random quantities, and thus statistical methods are used for various recognition tasks including decision making, feature extraction, learning, and cluster analysis. After 25 years of development, statistical pattern recognition has now been a well-established area[1] with a number of textbooks published[2-14] and a very large number of journal and conference publications. In this chapter, emphasis is placed on the statistical pattern recognition techniques that are suitable and effective for digital waveform recognition. Nonstatistical methods in pattern recognition, especially the syntactic and structural approaches[15,16] have in recent years gained much attention. Chapter 5 has a detailed discussion of syntactic pattern recognition with applications to waveform processing.

A digital waveform recognition system receives a digitized waveform as input, evaluates a feature set for the input measurement, and then classifies the waveform, based on the feature values, as belonging to one of several predetermined categories (pattern classes). Correct classification is the main objective in pattern recognition. Decision rules should be carefully chosen and utilized for minimum error classification. A feature set that has the largest discrimination power must be constructed from all features available. Parameters must be estimated or "learned" from the samples for classification and feature selection. Although pattern classes need not be defined in cluster analysis, it is desirable to assign all measurements to proper clusters according to certain optimum criteria. In addition to the topics stated above, special problem areas such as the finite sample size effect, recognition based on the incomplete data, etc., will also be discussed in this chapter.

II. DECISION RULES FOR WAVEFORM RECOGNITION

We start with the Bayes decision rule. Let x be a vector measurement of an input waveform and m be the number of classes. The Bayes decision rule minimizes the average risk with respect to the given *a priori* probabilities P_i, i = 1, 2, ---, m. For equal loss functions, the Bayes decision rule reduces to the maximum likelihood decision rule (MLDR), which chooses the class that maximizes the function

$$P_i p \ (x/\omega_i); \quad i = 1, 2, ---, m$$

where the conditional probability densities $p \ (x/\omega_i)$ must be known or estimated. Decision boundaries can be established for each class based on the specified decision rule. Let Ω_i be the measurement space of the ith class defined by the decision boundaries. Then the average error probability is given by

$$P_e = 1 - \sum_{i=1}^{m} P_i \int_{\Omega_i} p \ (x/\omega_i) \ dx \qquad (1)$$

For two multivariate Gaussian densities with mean μ_i and covariance matrix Σ_i, i = 1,2, the MLDR is to assign x to the class for which

$$(x - \mu_i)' \sum_i^{-1} (x - \mu_i) - \log\left(P_i^{\,2} / \left|\sum_i\right|\right) \qquad (2)$$

is the minimum. Assume further that $\Sigma_i = \Sigma_2 = \Sigma$. Then the decision boundary between the two classes is given by[11]

$$U(x) = C \tag{3}$$

where

$$U(x) = x' \sum^{-1} (\mu_1 - \mu_2) - \frac{1}{2} (\mu_1 + \mu_2)' \sum^{-1} (\mu_1 + \mu_2)$$

and

$$C = \log(P_2/P_1) \tag{4}$$

A given input measurement x is classified as belonging to Class 1 if $U(x) > C$, and Class 2 if $U(x) < C$. The error probability based on this decision rule is given by

$$P_e = \int_{(c + \delta/2)/\sqrt{\delta}}^{\infty} \frac{1}{\sqrt{2\pi}} \exp{-\frac{y^2}{2}} dy \tag{5}$$

since in this case the integrals in Equation 1 are the same. Here

$$\delta = (\mu_1 - \mu_2)' \sum^{-1} (\mu_1 - \mu_2) \tag{6}$$

which is known as the Mahalanobis distance between the two pattern classes. If $\Sigma_1 \neq \Sigma_2$, a simple but suboptimal procedure is to set $\Sigma = \frac{1}{2} (\mu_1 + \mu_2)$ and still use the decision rule given by Equations 3 and 4. In some applications[17] where $\Sigma_1 \neq \Sigma_2$, a modified MLDR that replaces Equation 2 by choosing the minimum of the form

$$(x - \mu_i)' \sum_i^{-1} (x - \mu_i) \tag{7}$$

can perform better than the MLDR. This is an example of the gap between theory and practice. The performance of the Bayes decision rule or the Bayes error probability given by Equation 1 in general cannot be expressed with a closed form. The error estimate, which critically depends on the sample size, is by itself a fundamental problem in statistics.[18]

The nearest neighbor decision rule (NNDR) identifies the vector measurement x with the class of its nearest neighbor; nearness being measured by the Euclidean distance or other suitably defined distance. For $k - NNDR$, the decision is based on the majority vote of k nearest neighbors. Let R_n be the risk incurred when the classification is based on the NNDR and n learning (training) samples are available. Also let R^* be the Bayes risk when the statistics are known exactly. As $n \to \infty$, R_n is tightly upper and lower bounded as[19]

$$R^* \leqslant R_n \leqslant R^* \left(2 - \frac{m}{m-1} R^*\right) \tag{8}$$

For m = 2,

$$R^* \leqslant R_n \leqslant 2R^* (1 - R^*) \leqslant 2R^* \tag{9}$$

Thus for any number of pattern classes, the risk (or the error probability) of the NNDR is bounded above by twice the Bayes risk (or error probability). The NNDR is nonparametric because the information on the probability densities is not needed. Obvious drawbacks of the NNDR are the large number of distance computation and the storage requirement of learning samples. Procedures to reduce the computation and storage requirements include the condensed NNDR,[20] edited NNDR,[21] selection of training samples,[22] and the use of branch and bound algorithms,[23] etc. By working with the preprocessed data, the branch and bound method requires less than 10% of distance computation normally required. Other modification of the NNDR uses the distance weighted NNDR, which can provide better recognition result in practice than the unweighted NNDR discussed above.[24]

The reject option has been considered for both the Bayes decision rule and the NNDR. The errors can be reduced at the expense of some rejects. The error-reject trade-off is an additional consideration in the reject option.

Linear, piecewise linear, and quadratic discriminant functions have been extensively investigated, especially in the statistical literatures. However, the closed form error probability expressions are generally unavailable except in the simple cases such as the multivariate Gaussian with unequal mean but equal covariance, as described above. The use of the MLDR is implied for the parametric discriminant analysis, and the optimization criterion is the minimum error probability. The Fisher's linear discriminant is a nonparametric technique that maximizes the ratio of between-class scatter to within-class scatter in the one-dimensional space on which the vector measurements are projected.[9] Let

$$\mu_1, \mu_2, \text{ and } \sum = \frac{1}{2}\left(\sum_1 + \sum_2\right) \tag{10}$$

be estimated from learning samples. Define

$$\alpha = \sum^{-1} (\mu_1 - \mu_2) \tag{11}$$

When a new vector measurement x is to be classified, we calculate

$$y = \alpha' x = (\mu_1 - \mu_2)' \sum^{-1} x \tag{12}$$

which is called the Fisher's linear discriminant. Its value is compared with a predetermined threshold, and x is classified as belonging to Class 1 if α' x is greater than the threshold value.

If $\Sigma_1 = \Sigma_2 = \Sigma$, we may choose the threshold as equal to $\frac{1}{2}(\mu_1 + \mu_2)' \Sigma^{-1} (\mu_1 - \mu_2)$, which leads to the same decision rule in connection with Equation 4. Note that Gaussian assumption of vector measurements is not made in using the Fisher's linear discriminant. If $\Sigma_1 \neq \Sigma_2$, a possible choice of the threshold value is[13]

$$\frac{1}{\left(\alpha'\sum_1\alpha\right)^{1/2} + \left(\alpha'\sum_2\alpha\right)^{1/2}} \left[\left(\alpha'\sum_1\alpha\right)^{1/2}\alpha'\mu_1 + \left(\alpha'\sum_2\alpha\right)^{1/2}\alpha'\mu_2\right]$$

Another major decision rule is the class of tree classifiers in which multistage decisions are made before a final decision is reached. For waveforms such as the speech and biomedical data, the tree classifiers may be needed to incorporate successively various subcategories, making the overall decision more reliable and even faster than the single stage classifiers discussed above. At each stage of the decision tree, a linear classifier may be used to reduce the computational complexity. The same or different sets of features may be used at each stage of the decision tree. The problem of optimum tree classifier design remains to be extensively studied.[25] The methods of reducing the computational complexity considered include clustering the decision rules, using branch and bound procedure to find efficient decision rule, and for feature assignment, etc. The decision tree classifier is the most promising classification mechanism for increasingly complex recognition problems.

Other generalizations of the conventional decision theory are the compound decision rules[11] and the multimembership classification, which allows simultaneous membership of a measurement in several classes.

III. STATISTICAL FEATURE EXTRACTION

The mathematical features as well as the structural features are best suited for automatic recognition, although they may not necessarily have physical meaning or may be quite different from features derived by human recognition process. A fundamental approach to extract features in statistical pattern recognition is by evaluating a number of available features to select a small subset of good features. Such evaluation can be based on the direct estimates of the error probability or an indirect measure of feature effectiveness as provided by the class of distance and information measures.[26,27] The distance measure selects features that have the largest statistical difference as measured by "distance", while the information measure selects features that have the most uncertainty or information. The relative effectiveness of various measures has been considered.[11] These measures are also very useful as error bounds.[28] A few important measures are listed below.

1. The divergence defined as

$$J = \int [P_1 \, p(x/w_1) - P_2 \, p(x/w_2)] \, \log \frac{p(x/w_1)}{p(x/w_2)} \, dx \qquad (13)$$

2. The Bhattacharyya distance defined as

$$b = -\log \rho \qquad (14)$$

where

$$\rho = \int \sqrt{P(x/w_1) \, p(x/w_2) \, dx}$$

is called the Bhattacharyya coefficient.

3. The mutual information defined as

$$I = \sum_{i=1}^{m} \int P_i \, p(x/w_i) \, \log \frac{P(w_i/x)}{P_i} \, dx \qquad (15)$$

Another useful approach is the linear orthogonal transformation methods.[5] If a pattern can be completely described by the second order statistics, the Karhunen-Loeve transform is optimal in the mean-square-error sense. The Karhunen-Loeve transform does require a considerable amount of computations. Other transforms, such as Fourier, Walsh, and discrete cosine, can be computed by fast algorithms. A large number of transform coefficients, however, are needed to form an effective feature set.

A realistic solution to feature extraction must take into account the nature of patterns, the *a priori* knowledge available, and the specific requirements and constraints of the given recognition task. Although an exhaustive search is about the only way to find the best feature set, search procedures have been considered[29] to provide a computationally feasible solution.

We now consider a practical example[13] of extracting features from the ECG (electrocardiogram) and VCG (vectorcardiogram). The mathematical approach would suggest to represent the signals

$$x_j(t) = \sum_{i=1}^{N} C_{ij}\ \phi_i(t) \tag{16}$$

where $x_j(t)$ is the jth lead of the ECG or VCG, $\phi_i(t)$ is one of a family of N suitable basis functions, and C_{ij} is for the coefficients that form a feature set. All of the systems that have been applied clinically, however, have used features that can be measured directly from the waveform. The simplest choice for a set of features is to measure the magnitude of the waveform at selected times in the heartbeat. A typical choice may use 15 equally spaced points in the QRS complex of each of the three VCG leads plus the duration of the QRS complex so that there are 46 features. Features can also be obtained from the ratios of selected amplitude measurements and from the time duration of the complete heartbeat and of various complexes (P, Q, R, S, and T) in the heartbeat. A large number of features usually result. In this case, simple linear decision functions may be used for classification. It is often necessary to reduce the feature space by mathematical and experimental procedures so that the total number of features is reduced to a manageable size of, say, 15 or lower.

We would like to emphasize that a real challenge to human intelligence is to extract the right features that truly characterize a pattern. Although much has been studied, feature extraction will remain to be a key problem in pattern recognition. Experimental methods should be relied on whenever the theoretical mechanism is inadequate.

IV. LEARNING IN PATTERN RECOGNITION

As the statistical information required for pattern classification is either partially or completely unavailable, a "learning" or estimation process is needed to obtain such information from the samples available. By following the statistical framework, learning in pattern recognition consists of the parametric learning and the nonparametric learning. In the parametric case, the probability densities required for decision making are assumed known except for some parameters that may be estimated by using Bayes, maximum likelihood, or other estimation methods. To estimate the parameters of certain class, it is usually assumed that samples belonging to that class are available for estimation purpose. This is the familiar supervised learning problem. The samples are called training or learning samples. In some cases, it is not known to which class the learning samples belong. Estimation is still possible from the available samples, and the problem known as unsupervised learning in statistical pattern recognition is one of identification of mixture parameters in mathematical statistics. The most important problem in nonparametric learning is the nonparametric probability density estimation, another subject extensively studied by statisticians. Only a general discussion of the learning procedures described above is given in this section. In-depth studies of machine learning in pattern recognition are available.[2-14]

The maximum likelihood estimate of the parameter θ is to maximize

$$P_i \, p \, (x/\theta, w_i) \tag{17A}$$

with respect to θ. The result is a set of equations from which θ can be obtained. The Bayes estimation procedure, however, assumes that θ is random, with known probability density $p \, (\theta)$. The Bayes estimate is the conditional expectation given by

$$\hat{\theta} = \int \theta \, p \, (\theta \, | x) \, d\theta \tag{17B}$$

It is noted that x in Equation 17 refers to the learning samples of the ith class. For the multivariate Gaussian density with its logarithm given by Equation 2, the maximum likelihood estimate of the mean vector is given by

$$\hat{\mu} = \frac{1}{N} \sum_{i=1}^{N} x_i \tag{18A}$$

where χ_i, $i = 1,2,...,N$ is the learning sample. The covariance matrix has the maximum likelihood estimate

$$\hat{\Sigma} = \frac{1}{N-1} \sum_{i=1}^{N} (x_i - \hat{\mu}) \, (x_i - \hat{\mu})' \tag{18B}$$

To obtain the Bayes estimate, we assume that Σ is known but μ is random with Gaussian distribution of mean and covariance given by μ_o and ϕ_o, respectively. Then the Bayes estimate based on N samples is

$$\hat{\mu}_N = \frac{1}{N}\Sigma\left(\phi_0 + \frac{1}{N}\Sigma\right)^{-1} \mu_0 + \phi_0 \left(\phi_0 + \frac{1}{N}\Sigma\right)^{-1} \frac{1}{N} \sum_{i=1}^{N} x_i$$

$$\phi_N = \frac{1}{N}\Sigma\left(\phi_0 + \frac{1}{N}\Sigma\right)^{-1} \phi_0 \tag{19}$$

As $N \to \infty \ \phi_N \to o$ and $\hat{\mu}_N \to \hat{\mu}$ given by Equation 18. Equations 18A and 19 can also be written in recursive forms. In general, the estimated parameters can be updated with every new input measurement. Presumably the word "learning" implies performance improvement as the measurements are taken sequentially. The recursive relation holds, however, only in certain cases. For the Bayes estimate, this requires that the probability density of the parameters be "reproducing" with respect to the probability density of the measurements.[9]

To illustrate the unsupervised learning, consider again the Gaussian densities with one dimension only. Although the samples are not classified, each sample belongs to the mixture of probability densities

$$p(x) = \sum_{i=1}^{m} P_i \, p(x/w_i) \qquad (20)$$

which is generally a multimodal distribution. For two classes, the probability densities are

$$p(x/w_i) = \frac{1}{\sqrt{2\pi}\, \sigma_0} \exp - \frac{(x - \mu_i)^2}{2\sigma_0^2}; \; i = 1,2 \qquad (21)$$

where σ_o is assumed as known but μ_i is unknown. Let $\hat{\mu}$ and $\hat{\sigma}^2$ be the estimated mean and variance, respectively, of the mixture density $p(x)$ obtained from unclassified samples. Then μ_i, $i = 1,2$, the estimate of μ_i, can be solved from the following two equations.

$$\hat{\mu} = P_1 \hat{\mu}_1 + P_2 \hat{\mu}_2$$

$$\hat{\sigma}^2 + \hat{\mu}^2 = P_1 (\sigma_0^2 + \hat{\mu}_1^2) + P_2 (\sigma_0^2 + \hat{\mu}_2^2) \qquad (22)$$

By the same procedures, parameters of the component densities $p(x/w_i)$ can be determined by using the estimated moments of the mixture density. Equation 22 is representative of the maximum likelihood estimation method for unsupervised learning. The Bayesian procedure for unsupervised learning is more suitable for communications problems.[11]

For the nonparametric learning, we are concerned with the nonparametric probability density estimation. Major approaches include the kernel estimators, notably the Parzen estimator, the k-nearest neighbor estimate, the classical orthogonal series estimation, and the histogram estimator and its modifications.

In the histogram approach we partition the measurement space into a number of mutually disjoint cells. The probability density is approximated by the number of samples in each cell. The histogram estimator is clearly distribution free and has the advantage over the classical series estimator of being more local. The cell size can be fixed or variable. The disadvantages of the histogram estimator are that it is discontinuous, completely *ad hoc*, and requires large storage space.

Let x_i, $i = 1, ..., N$ be continuous one-dimensional random measurements with unknown density $p(x)$. An important modification of the histogram estimator is called the Rosenblatt estimator given by[30,31]

$$\hat{p}_N(x) = \frac{\text{\# measurements in } (x - h_N, \, x + h_N)}{2Nh_N} \qquad (23A)$$

where h_N is a real valued constant for each N, i.e.,

$$\hat{p}_N(x) = \frac{P_N(x + h_N) - P_N(x - h_N)}{2h_N} \qquad (23B)$$

where

$$P_N(x) = \frac{\text{\# measurements} \leqslant x}{N} \qquad (23C)$$

is the empirical cumulative distribution function.

In comparison with the histogram estimator, the Rosenblatt kernel estimator is simply a histogram that, for estimating the density at x, has been shifted so that x lies at the center of the cell interval. Note that the bias of the resulting estimator is reduced by the shifted histogram. The shifted histogram has another representation given by

$$\hat{p}_N (x) = \frac{1}{N} \sum_{j=1}^{N} \frac{1}{h_n} W \left(\frac{x - x_j}{h_n} \right) \tag{24}$$

where W(u) is a rectangular "window" defined as

$$W(u) = \frac{1}{2} \text{ if } |u| < 1$$

$$= 0$$

otherwise. It is clear that $\int \hat{p}_N(x) \, dx = 1$ and $\hat{p}_N(x) \geqslant 0$ and thus $\hat{p}_N(x)$ satisfies the condition for probability density. A more general representation of the kernel estimators, due to Parzen,[32] is given by

$$p_N (x) = \int_{-\infty}^{\infty} \frac{1}{h_N} K \left(\frac{x-y}{h_N} \right) dP_N (y) = \frac{1}{Nh_N} \sum_{j=1}^{N} K \left(\frac{x - x_j}{h_N} \right) \tag{25}$$

where

$$\int_{-\infty}^{\infty} |K(y)| dy < \infty$$

$$\sup_{-\infty < y < \infty} |K(y)| < \infty$$

$$\lim_{y \to \infty} |yK(y)| = 0 \tag{26}$$

$$K(y) \geqslant 0$$

and

$$\int_{-\infty}^{\infty} K(y) \, dy = 1$$

It can be shown that the kernel estimator $\hat{p}_N$ in Equation 25 subject to Equation 26 is asymptotically unbiased if $h_N \to 0$ as $N \to \infty$, i.e.,

$$\lim E (\hat{p}_N(x)) = \rho (x)$$

and furthermore $\hat{p}_N$ is consistent if we add the additional constraint $\lim N h_N \to \infty$.

By using maximum likelihood approach, Wegman[33] obtained a modified histogram estimator with interval widths of the histogram varying across the data base in a manner inversely proportional to the density of measurements in the interval. Another recent approach[34] motivated by the estimation of spectral densities of second order stationary time series is to approximate the probability density by autoregressive model.

The classical series estimator represents the probability density by orthonormal series.[11] In the k-nearest neighbor approach, the probability density estimate is given by[7]

$$\hat{p}_N(x) = \frac{k-1}{N} \frac{1}{A(k,N,x)} \tag{27}$$

where k is the number of nearest neighbors considered, $A(k,N,x)$ is the volume (or length in the one-dimensional measurement case) of the hypersphere for the set of all points whose distance from x is less than γ. Here, γ is chosen to include k-nearest neighbors. The k-nearest neighbors estimation provides a very simple density estimate. However, γ should be small enough to keep density function relatively constant within the hypersphere. Thus, k would have to be small to sacrifice the accuracy.

As a concluding remark, the parametric assumption of probability density of a digital waveform is often unsuitable. The nonparametric kernel estimators described above are very useful for the much-needed probability density estimation from the available measurements in waveform recognition study.

V. CLUSTER ANALYSIS

The clustering problem is to find natural groupings of a set of data. In unsupervised learning, we are concerned with classifying input samples by using the statistical information extracted from the unclassified learning samples. In cluster analysis both the statistical and the structural description of the data are considered in addition to the unsupervised learning problem. Similarity and distance measures are very useful to assign each sample to an appropriate cluster. Let x_i and x_j be two N-dimensional vector measurements. Typical similarity measures are

$$S_{ij} = \frac{x_i' x_j}{\sqrt{(x_i' x_i)\ (x_j' x_j)}} \tag{28}$$

which is a normalized correlation, and

$$S_{ij} = \frac{x_i' x_j}{x_i' x_i + x_j' x_j - x_i' x_j} \tag{29}$$

which is frequently used in information retrieval and biological taxonomy. The Eucliden distance

$$d_{ij}^2 = (x_i - x_j)'(x_i - x_j) = \sum_{k=1}^{N} (x_{ik} - x_{jk})^2 \tag{30}$$

and the weighted distance

$$D_{ij}^2 = \sum_{k=1}^{N} w_k (x_{ik} - x_{jk})^2 \tag{31}$$

where w_k is a weighting function, are typical distance measures that measure the dissimilarity between x_i and x_j. Clusters defined by Euclidean distances are invariant to translation or rotation of coordinates. However, they are not invariant to transformations that change the distance relationships.

Many data sets have certain structural properties. The structural information should be used in cluster analysis. Nonstatistical procedures, such as the graph theoretic methods and the hierarchical clustering, are very effective for such data sets.

In statistical cluster analysis, the problem is to obtain a partition that extremizes some criterion function. Let x_i, x_2,---,x_n be n M-dimensional samples that are partitioned into M groups, G_1, G_2, ---, G_M, with n_1, n_2, ---, n_m samples, respectively. Let n $= n_1 + n_2 + \cdots + n_M$. Define scatter matrices as follows:

Total scatter:

$$T = \sum_{k=1}^{n} x_k x_k{}'$$

Intergroup scatter:

$$W_j = \sum_{x_k \epsilon G_j} (x_k - C_j)\ (x_k - C_j)'$$

where C_j is the sample mean of G_j

$$G_j,\ C_j = \frac{1}{n_j} \sum_{x_k \epsilon G_j} x_k$$

Total intragroup scatter:

$$W = \sum_{j=1}^{M} W_j$$

Intergroup scatter:

$$B = \sum_{j=1}^{M} n_j C_j C_j'$$

It can easily be shown that $T = W + B$, regardless of the partition. One criterion function is defined as the sum of mean square distances to the group centers

$$J_0 = \sum_{j=1}^{M} \sum_{x_k \epsilon G_j} |x_k - C_j|^2 = \text{tr } W \qquad (32)$$

where tr W is the trace of the matrix W. The optimum partition is taken as the one that minimizes J_o. Clusterings of this type are often called minimum variance partitions. J_o is not invariant under nonsingular linear transformation of the x_k's. This means that by changing the coordinate system of the original data, one may alter the optimum partition. It is not difficult to show that the eigenvalues $\lambda_1, \lambda_2, ---, \lambda_N$ of $W^{-1}B$ are invariant under nonsingular linear transformations of the data. Since the trace of a matrix is the sum of its eigenvalues, an invariant criterion function defined as

$$J_1 = \text{tr } W^{-1} B = \sum_{i=1}^{N} \lambda_i \qquad (33)$$

can be maximized for an optimum partition. A similar invariant criterion function is

$$J_2 = \left| \frac{T}{W} \right| = \prod_{i=1}^{N} (1 + \lambda_i) \qquad (34)$$

The above criterion functions are more suitable for data sets where various clusters are reasonably well-separated. For example, they will not extract a very dense cluster embedded in the center of a diffuse cluster. Also, the invariant criterion functions given by Equations 33 and 34 are essentially similar.

Once the criterion function has been selected, clustering becomes a problem of iterative optimization, which seeks for those partitions of the data set that extremize the criterion function. The ISODATA (an iterative self-organizing data analysis)[35] and k-means procedures[36] are two examples of iterative optimization for clustering. The number of clusters need not be known in the ISODATA. The method starts with initial cluster centers, which are modified after splitting or lumping of clusters at each iteration. The k-means algorithm also starts with initial cluster centers, but the number of clusters remains unchanged. A new cluster center is computed at each iteration by taking the sample mean of samples belonging to the cluster. Samples are assigned to a cluster with the smallest Euclidean distance. When there is no change in the sequential update of cluster centers, the algorithm has converged and the procedure stops.

Another important and broad problem area in cluster analysis is the nonlinear mapping of high-dimensional data to a low-dimensional space without changing the data structure. This topic, also called multidimensional scaling, is particularly important for data structure analysis in two-dimensional space. It has many applications in psychology and social sciences, in addition to engineering and physical sciences. As the cluster analysis usually depends on a set of parameters that can be adjusted by trial and error, interactive procedures are very useful for clustering study. Extensive references are available[11] for topics discussed in this section.

VI. RECOGNITION OF INCOMPLETE VECTORS

Vector measurements taken from a digitized waveform with missing data form incomplete vectors for pattern recognition. Occasional missing samples, well-separated from each other, can be estimated with simple interpolation procedures (polynomial or parabolic fits, spline fits, etc.). However, data may be missing in such quantities that conventional interpolation is inadequate; data gaps longer than the periods of the sinusoidal components in the data cannot be easily bridged with simple functions. A more sophisticated approach becomes necessary, and the use of data-adaptive linear prediction filter[37] is one feasible alternative.

Missing data in waveforms may occur for a variety of reasons. An intercepted radar pulse may be so weak and noisy that some features are not measurable. Hardware may fail to transmit or receive signals properly. The method of collecting data may change during the course of an investigation so that a feature is measured in some cases but not in others. Whatever the cause may be, recognition of incomplete vectors presents an important practical problem. A recognition system should be capable of accepting imperfect data while still performing reasonably well.

Sebestyen[2] gives a theoretical analysis of pattern recognition with partially missing data. He assumes that the conditional probability of class membership given any set or partial set of measurements is known. Four decision rules were considered: decision based on the marginal densities of the actually observed variables; decision rule that predicts the most likely values of the missing measurements and uses them as if they had actually been measured; decision using most probable value of the likelihood ratio; and decision using average value of likelihood ratio. His conclusion is that one should use only the actually measured values; no useful purpose is served by attempting to estimate the missing values. Indeed, extrapolation to a higher, but unavailable, dimensional space is not feasible. If the probability density is unknown, nonparametric density estimation and orthogonal series expansion[38] as well as the use of the discriminant function based on the minimum mean squared error criterion[39] have all been considered. In digital waveforms, the missing data occur randomly and affect several or all features in a pattern vector. Good prediction procedure for the missing data is desirable or even necessary. After data prediction, cluster methods[40] can be used to compute the distance between two vectors and then to normalize to further adjust the predicted values. If data prediction is not performed, then the distance to a missing feature can be set as the average distance between nonmissing features. Without data prediction, the features computed are less reliable. Features may also be extracted from the frequency domain as the averaging operations in frequency transform tend to reduce the effect of errors in the time domain.

VII. FINITE SAMPLE AND FEATURE SIZES

In most digital waveform recognition problems, the number of samples available for classification and for learning or parameter estimation are limited. They are called the finite sample size problems. The feature size or dimensionality is always finite in practice, as only a limited number of features are extracted for classification. There are several interesting phenomena resulting from the finite sample and feature sizes. These are long-standing problems in statistical decision and estimation theory, which are fundamental to the pattern recognition studies.

For a given number of design or learning samples that are used to estimate parameters by the maximum likelihood estimation procedure, increasing the dimensionality first increases the probability of correct recognition until it reaches certain peak value beyond which further increase in the dimensionality will actually decrease the probability of correct recognition. This is the well-known peaking phenomenon due to the finite sample size. The peaks are not sharp and will disappear as the sample size increases. If the parameters and thus the probability densities are completely known, the peaking phenomenon does not exist.

A simple example was considered by Trunk[41] who assumed two multivariate Gaussian densities with means $\mu_1 = \mu$, $\mu_2 = -\mu$, and identical covariance matrices $\Sigma_1 = \Sigma_2 = I$. If μ is known, then the error probability is

$$P_e = \int_{r/2}^{\infty} \frac{1}{\sqrt{2\pi}} e^{-\frac{y^2}{2}} \, dy \qquad (35)$$

where $r^2 = |\mu^1 - \mu_2|^2 = 4 \sum\limits_{i=1}^{n} (1/i)$. As $n \to \infty$, $r^2 \to \infty$ and thus $P_e \to 0$. If μ is estimated by the sample mean from N samples of known classification $X_1, ---, X_N$, i.e.,

$$\hat{\mu} = \frac{1}{N} \sum_{i=1}^{N} X_i \tag{36}$$

where X_i has been replaced by $-X_i$ if X_i came from Class 2. The error probability can be shown as

$$P_e = \int_{\gamma_n}^{\infty} \frac{1}{\sqrt{2\pi}} e^{-\frac{y^2}{2}} dy$$

where $\gamma_n = E(z)/[Var(z)]^{1/2}$, $z = X' \hat{\mu}$ and

$$E(z) = \sum_{i=1}^{n} (1/i)$$

$$Var(z) = (1 + \frac{1}{N}) \sum_{i=1}^{n} (1/i) + n/N$$

It can be shown that

$$\lim_{n \to \infty} \gamma_n = 0 \tag{37}$$

and thus the error probability approaches one half as the dimensionality becomes very large. Typical set of optimum dimensionality n_{opt}. can be listed as follows:

N =	1	4	10	25	100	∞
n_{opt} =	5	10	30	70	200	∞

It is important to note that the above example merely illustrates the existence of peaking phenomenon. For practical purposes, the sample size should be several times of the feature size,[42] a condition that is not difficult to meet in practice, for the best recognition results.

The peaking phenomena depend very much on the probability structures and the decision rules and estimation procedures employed.[43] Mean or average recognition performance vs. the finite sample and feature sizes has also been examined extensively.[44] Unfortunately, theoretical results based on different assumptions are inconsistent and must be used with care. The best rule for designing a waveform recognition system is to use a small number of effective features and a sufficient number of design samples, at least several times the feature size.

Another finite sample problem is the estimation of error probability when the total number of samples for design and testing is fixed. A recommended solution is the "hold-one-out" method.[45-47] In this method, a classifier is designed on N−1 samples and tested on the one remaining sample. This procedure is then repeated for all N samples. The number of errors counted provides an estimate of the error probability. For other problems of the finite learning samples, see References 48 and 49.

VIII. CONCLUDING REMARKS

In this chapter, we have presented a basic theory of statistical pattern recognition, especially with waveform recognition in mind. Although there are still many theoretically unresolved problems in statistical pattern recognition, a good waveform recognition system may be designed according to the existing theory. For unusual requirements in recognition performance, the present theory is inadequate. In this case, other techniques such as interactive systems, structural pattern recognition, etc., can be helpful. More research is needed on the use of mixed approaches, which appear to be necessary as the patterns studied become more complex and the recognition requirements become more stringent.

REFERENCES

1. Chen, C. H., A review of statistical pattern recognition, in *Pattern Recognition and Artificial Intelligence*, Chen, C. H., Ed., Sitjhoff & Noordhoff International Publishing, Winchester, Mass., NATO ASI Series E, No. 29, 1978.
2. Sebestyen, G. S., *Decision-Making Processes in Pattern Recognition*, Macmillan, New York, 1962.
3. Nilsson, N. J., *Learning Machines-Foundations of Trainable Pattern-Classifying Systems*, McGraw-Hill, New York, 1965.
4. Fu, K. S., *Sequential Methods in Pattern Recognition and Machine Learning*, Academic Press, New York, 1968.
5. Andrews, H. C., *Introduction to Mathematical Techniques in Pattern Recognition*, John Wiley & Sons, New York, 1972.
6. Meisel, W. S., *Computer-Oriented Approaches to Pattern Recognition*, Academic Press, New York, 1972.
7. Fukunaga, K., *Introduction to Statistical Pattern Recognition*, Academic Press, New York, 1972.
8. Patrick, E. A., *Fundamentals of Pattern Recognition*, Prentice-Hall, Englewood Cliffs, N.J., 1972.
9. Duda, R. O. and Hart, P. E., *Pattern Classification and Scene Analysis*, John Wiley & Sons, New York, 1973.
10. Ullmann, J. R., *Pattern Recognition Techniques*, Butterworths, London, 1973.
11. Chen, C. H., *Statistical Pattern Recognition*, Hayden Book, Rochelle Park, N.J., 1973.
12. Tou, J. T. and Gonzalez, R. C., *Pattern Recognition Principles*, Addison-Wesley, Reading, Mass., 1974.
13. Young, T. Y. and Calvert, T. W., *Classification, Estimation and Pattern Recognition*, American Elsevier, New York, 1974.
14. Batchelor, B. G., *Practical Approach to Pattern Classification*, Plenum Press, New York, 1974.
15. Fu, K. S., *Syntactic Methods in Pattern Recognition*, Academic Press, New York, 1974.
16. Pavlidis, T., *Structural Pattern Recognition*, Srpinger-Verlag, Basel, 1977.
17. Chang, J. K., Modified maximum likelihood decision rule and minimax Bayes decision rule, *Proc. 3rd Int. Jt. Conf. on Pattern Recognition*, IEEE, Piscataway, N.J., 1976.
18. Toussaint, G. T., Bibliography on estimation of misclassification, *IEEE Trans. Inf. Theory*, IT-20, 472, 1974.
19. Cover, T. M. and Hart, P. E., Nearest neighbor pattern classification, *IEEE Trans. Inf. Theory*, IT-13, 21, 1967.
20. Hart, P. E., The condensed nearest neighbor rule, *IEEE Trans. Inf. Theory*, IT-14, 50, 1968.
21. Wilson, D. L., Asymptotic properties of nearest neighbor rules using edited data, *IEEE Trans. Syst. Man Cybern.*, SMC-2(3), 408, 1972.

22. **Chen, C. H.,** Seismic pattern recognition, *Geoexploration J.,* 16(½), 133, April 1978.

23. **Fukunaga, K. and Narendra, P. M.,** A branch and bound algorithm for computing k-Nearest neighbors, *IEEE Trans. Comput.,* C-24, 750, 1975.

24. **Dudani, S. A.,** The distance-weighted k-nearest neighbor rule, *IEEE Trans. Syst. Man Cybern.,* SMC-6, 325, 1976.

25. **Kulkarni, A. V. and Kanal, L. N.,** An optimization approach to hierarchical classifier design, *Proc. 3rd Int. Conf. on Pattern Recognition,* IEEE, Piscataway, N.J., 1976, 459.

26. **Kanal, L. N.,** Patterns in pattern recognition, *IEEE Trans. Inf. Theory,* IT-20, 697, 1974.

27. **Chen, C. H.,** On information and distance measures, error bounds and feature selection, *Inf. Sci. J.,* 10, 159, 1976.

28. **Devijver, P. A.,** On a new class of bounds on Bayes risk in multihypothesis pattern recognition, *IEEE Trans. Comput.,* C-23, 70, 1974.

29. **Kittler, J.,** Feature set search algorithms, in *Pattern Recognition and Signal Processing,* Chen, C. H., Ed., Sitjhoff & Noordhoff, The Netherlands, 1978.

30. **Rosenblatt, M.,** Remarks on some nonparametric estimates of a density function, *Ann. Math. Stat.,* 27, 832, 1956.

31. **Tapia, R. A. and Thompson, J. R.,** *Nonparametric Probability Density Estimation,* The Johns Hopkins University Press, Baltimore, 1978.

32. **Parzen, E.,** On estimation of a probability density function and mode, *Ann. Math. Stat.,* 33, 1065, 1962.

33. **Wegman, E. J.,** Maximum likelihood estimation of a unimodal density function, *Ann. of Math. Stat.,* 41, 457, 1970.

34. **Carmichael, J.,** The Autoregressive Method: A Method of Approximating and Estimating Positive Functions, Ph.D. dissertation, State University of New York at Buffalo, 1976.

35. **Ball, G. H. and Hall, D. J.,** ISODATA, An Interative Method of Multivariate Data Analysis and Pattern Classification, Institute of Electrical and Electronics Engineers International Communication Conference Record, Philadelphia, June 1966.

36. **MacQueen, J.,** Some methods for classification and analysis of multivariate observations, *Proc. 5th Berkeley Symp. on Probability and Statistics,* University of California, Berkeley, 1967, 281.

37. **Bowling, S. B. and Lai, S.,** Use of Linear Prediction for the Interpolation and Extrapolation of Missing Data and Data Gaps, Report TN-1979-46, Lincoln Laboratory, Massachusetts Institute of Technology, Cambridge, 1979.

38. **Hand, D. J. and Batchelor, B. G.,** The classification of incomplete vectors, *Proc. 3rd Int. Jt. Conf. on Pattern Recognition,* IEEE, Piscataway, N.J., 1976.

39. **Kittler, J.,** Classification of incomplete pattern vectors using modified discriminant functions, *IEEE Trans. Comput.,* C-27(4), 367, 1978.

40. **Dixon, J. K.,** Pattern recognition with partly missing data, *IEEE Trans. on Syst. Man, Cybern.,* SMC-9(10), 617, 1979.

41. **Trunk, G. V.,** A problem of dimensionality: a simple example, *IEEE Trans. Pattern Anal. Machine Intelligence,* PAMI-1(3), 306, 1979.

42. **Foley, D. H.,** Considerations of sample and feature size, *IEEE Trans. Inf. Theory,* IT-18(5), 618, 1972.

43. **Raudys, S. J.,** Determination of optimal dimensionality in statistical pattern classification, *Pattern Recognition,* 11(4), 263, 1979.

44. **Hughes, G. F.,** On the mean accuracy of statistical pattern recognizers, *IEEE Trans. Inf. Theory,* 14, 55, 1968.

45. **Lachenbruch, P. A. and Mickey, M. R.,** Estimation of error rates in discriminant analysis, *Technometrics,* 10(1), February 1968.

46. **Kanal, L. and Chandrasekaram, B.,** On Dimensionality and Sample Size in Statistical Pattern Classification, in *Proc. 1968 Natl. Electron. Conf.,* National Electronic Conference, Oak Brook, Ill., 1968, 2.

47. **Toussaint, G. T.,** Bibliography on estimation of misclassification, *IEEE Trans. Inf. Theory,* IT-20(4), 472, 1974.

48. **Chen, C. H.,** Finite sample considerations in statistical pattern recognition, *Proc. Pattern Recognition and Image Processing Conf.,* IEEE, Piscataway, N.J., 1978.

49. **Pau, L. F.,** Finite learning sample size problems in pattern recognition, in *Pattern Recognition and Signal Processing,* Chen, C. H., Ed., Sijthoff & Noordhoff, The Netherlands, 1978.

Chapter 5

SYNTACTIC PATTERN RECOGNITION AND ITS APPLICATIONS TO SIGNAL PROCESSING*

K. S. Fu

TABLE OF CONTENTS

*   This work was supported by the National Science Foundation Grant ENG 78-16970 and the U.S. - Italy Cooperative Science Program.

I. INTRODUCTION

The many different mathematical techniques used to solve pattern recognition problems may be grouped into two general approaches.[1,2] They are the decision-theoretic (or discriminant) approach and the syntactic (or structural) approach.[3] In the decision-theoretic approach, a set of characteristic measurements, called features, are extracted from the patterns. Each pattern is represented by a feature vector, and the recognition of each pattern is usually made by partitioning the feature space. On the other hand, in the syntactic approach, each pattern is expressed as a composition of its components, called subpatterns and pattern primitives. This approach draws an analogy between the structure of patterns and the syntax of a langauge. The recognition of each pattern is usually made by parsing the pattern structure according to a given set of syntax rules. In this paper, we briefly review the recent progress in syntactic pattern recognition and some of its applications.

A block diagram of a syntactic pattern recognition system is shown in Figure 1. We divide the block diagram into the recognition part and the analysis part, where the recognition part consists of preprocessing, primitive extraction (including relations among primitives and subpatterns), and syntax (or structural) analysis, and the analysis part includes primitive selection and grammatical (or structural) inference.

In syntactic methods, a pattern is represented by a sentence in a language that is specified by a grammar. The language that provides the structural description of patterns, in terms of a set of pattern primitives and their composition relations, is sometimes called the "pattern description language". The rules governing the composition of primitives into patterns are specified by the so-called "pattern grammar". An alternative representation of the structural information of a pattern is to use a "relational graph", of which the nodes represent the subpatterns and the branches represent the relations between subpatterns.

Figure 2 gives an illustrative example for the description of the boundary of a submedian chromosome. The hierarchical structural description is shown in Figure 2A, and the grammar generating submedian chromosome boundaries is given in Figure 2B.

II. PRIMITIVE SELECTION AND PATTERN GRAMMARS

Since pattern primitives are the basic components of a pattern, presumably they are easy to recognize. Unfortunately, this is not necessarily the case in some practical applications. For example, strokes are considered good primitives for script handwriting, and so are phonemes for continuous speech; however, neither strokes nor phonemes can easily be extracted by machine. A compromise between its use as a basic part of the pattern and its easiness for recognition is often required in the process of selecting pattern primitives.

There is no general solution for the primitive selection problem at this time.[3-5] For line patterns or patterns described by boundaries or skeletons, line segments are often suggested as primitives. A straight line segment could be characterized by the locations of its beginning (tail) and end (head), its length, and/or slope. Similarly, a curve segment might be described in terms of its head and tail and its curvature. The information characterizing the primitives can be considered as their associated semantic information or as features used for primitive recognition. Through the structural description and the semantic specification of a pattern, the semantic information associated with its subpatterns or the pattern itself can then be determined. For pattern description in terms of regions, half-planes have been proposed as primitives. Shape and texture measurements are often used for the description of regions.

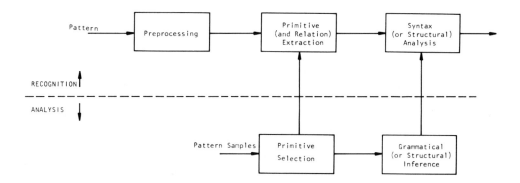

FIGURE 1. Block diagram of a syntactic pattern recognition system.

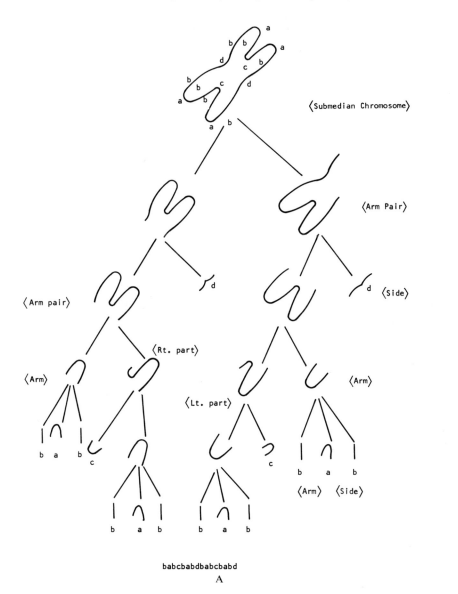

babcbabdbabcbabd

A

FIGURE 2. Syntactic representation of submedian chromosome.

$$G = (V_N, V_T, P, <\text{Submedian}>)$$

$$\text{where } V_N = \{<\text{Submedian}>, <\text{Arm pair}>, <\text{Lt. part}>,$$

$$<\text{Rt. part}>, <\text{Arm}>, <\text{Side}>\}$$

$$V_T = \{\overset{\cap}{a,} \mid b, \overset{\cup}{c,} \mid d\}$$

and P:

$<$ Submedian$>$	$\rightarrow$	$<$Arm pair$>$ $<$Arm pair$>$
$<$Arm pair$>$	$\rightarrow$	$<$Arm pair$>$ $<$Side$>$
$<$Arm pair$>$	$\rightarrow$	$<$Arm$>$ $<$Rt. part$>$
$<$Arm pair$>$	$\rightarrow$	$<$Lt. part$>$ $<$Arm$>$
$<$Rt. part$>$	$\rightarrow$	c$<$Arm$>$
$<$Lt. part$>$	$\rightarrow$	$<$Arm$>$c
$<$Arm$>$	$\rightarrow$	b$<$Arm$>$
$<$Arm$>$	$\rightarrow$	$<$Arm$>$b
$<$Side$>$	$\rightarrow$	b$<$Side$>$
$<$Side$>$	$\rightarrow$	$<$Side$>$b
$<$Arm$>$	$\rightarrow$	a
$<$Side$>$	$\rightarrow$	b
$<$Side$>$	$\rightarrow$	d

FIGURE 2B

After pattern primitives are selected, the next step is the construction of a grammar (or grammars) that will generate a language (or languages) to describe the patterns under study. It is known that increased descriptive power of a language is paid for in terms of increased complexity of the syntax analysis system (recognizer or acceptor). Finite-state automata are capable of recognizing finite-state languages, although the descriptive power of finite-state languages is also known to be weaker than that of context-free and context-sensitive languages. On the other hand, nonfinite, nondeterministic procedures are required, in general, to recognize languages generated by context-free and context-sensitive grammars. The selection of a particular grammar for pattern description is affected by the primitives selected and by the trade-off between the grammar's descriptive power and analysis efficiency.

If the primitives selected are very simple, more complex grammars may have to be used for pattern description. On the other hand, a use of sophisticated primitives may result in rather simple grammars for pattern description, which in turn will result in fast recognition algorithms. The interplay between the complexities of primitives and of pattern grammars is certainly very important in the design of a syntactic pattern recognition system. Context-free programmed grammars, which maintain the simplicity of context-free grammars but can generate context-sensitive languages, have recently been suggested for pattern description.[3]

A number of special languages have been proposed for the description of patterns such as English and Chinese characters, chromosome images, spark chamber pictures, two-dimensional mathematics, chemical structures, spoken words, and fingerprint patterns.[3,6] For the purpose of effectively describing high-dimensional patterns, high-dimensional grammars such as web grammars, graph grammars, tree grammars, and shape grammars have been used for syntactic pattern recognition.[3,7,8]

Ideally speaking, it would be nice to have a grammatical (or structural) inference machine that would infer a grammar from a given set of patterns. Unfortunately, not many convenient grammatical inference algorithms are presently available for this purpose.[9-15] Nevertheless, recent literatures have indicated that some simple grammatical inference algorithms have already been applied to syntactic pattern recognition, particularly through man-machine interaction.[16-19]

In some practical applications, a certain amount of uncertainty exists in the process under study. For example, due to the presence of noise and variation in the pattern measurements, segmentation error and primitive extraction error may occur, causing ambiguities in the pattern description languages. In order to describe noisy and distorted patterns under ambiguous situations, the use of stochastic languages has been suggested.[3] With probabilities associated with grammar rules, a stochastic grammar generates sentences with a probability distribution. The probability distribution of the sentences can be used to model the noisy situations.

A stochastic grammar is a quadruple $G_s = (V_N, V_T, P_s, S)$ where P_s is a finite set of stochastic productions. For a stochastic context-free grammar, a production in P_s is of the form

$$A_i \xrightarrow{P_{ij}} \alpha_j, A_i \in V_N, \alpha_j \in (V_N \cup V_T)^*$$

where P_{ij} is called the production probability. The probability of generating a string x, called the string probability p(x), is the product of all production probabilities associated with the productions used in the generation of x. The language generated by a stochastic grammar consists of the strings generated by the grammar and their associated string probabilities.

By associating probabilities with the strings, we can impose a probabilistic structure on the language to describe noisy patterns. The probability distribution characterizing the patterns in a class can be interpreted as the probability distribution associated with the strings in a language. Thus, statistical decision rules can be applied to the classification of a pattern under ambiguous situations (for example, use the maximum-likelihood or Bayes decision rule). A block diagram of such a recognition system using maximum-likelihood decision rule is shown in Figure 3. Furthermore, because of the availability of the information about production probabilities, the speed of syntactic analysis can be improved through the use of this information.[3] Of course, in practice, the production probabilities will have to be inferred from the observation of relatively large numbers of pattern samples.[3] When the imprecision and uncertainty involved in the pattern description can be modeled by using the fuzzy set theory, the use of fuzzy languages for syntactic pattern recognition has been suggested.[53]

Other approaches for the recognition of distorted or noisy patterns using syntactic methods include the use of transformational grammar[26,27] and approximation,[25] and the application of error-correcting parsing techniques.[24,28-31]

III. SYNTACTIC CLUSTERING

Errors in a string[30] are considered to be the following three types: substitution, dele-

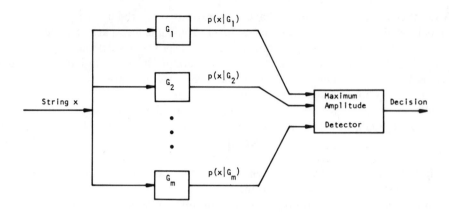

FIGURE 3.   Maximum-likelihood syntactic recognition system.

tion, and insertion of errors, which are treated as syntax errors by defining transformations from V_T^* to a subset of V_T^*.

Definition 1 — For two strings, x, y ε V_T^*, define a transformation T: $V_T^* \rightarrow V_T^*$ such that y ε T(x). T has the following three types:

1.   Substitution error transformation

$$\omega_1 a \omega_2 \ \vert \xrightarrow{\ T_\sigma\ } \ \omega_1 b \omega_2, \text{ for all a, b } \epsilon \ V_T, a \neq b$$

2.   Deletion error transformation

$$\omega_1 a \omega_2 \ \vert \xrightarrow{\ T_\epsilon\ } \ \omega_1 \omega_2, \text{ for all a } \epsilon \ V_T$$

3.   Insertion error transformation

$$\omega_1 \omega_2 \ \vert \xrightarrow{\ T_\phi\ } \ \omega_1 a \omega_2, \text{ for all a } \epsilon \ V_T$$

where ω_1, ω_2 ε V_T^* are substrings of x.

Definition 2 — The distance between two strings x, y, d(x, y), is defined as the smallest number of transformations required to derive y from x.

Example 1 — Given a sentence x = cbabdbb and a sentence y = cbbabbdb, then, x = cbabdbb

$$\vert \xrightarrow{\ T_\sigma\ } \text{ cbabbbb } \vert \xrightarrow{\ T_\sigma\ } \text{ cbabbdb } \vert \xrightarrow{\ T_\phi\ } \text{ cbbabbdb } = y$$

The minimum number of transformations required to transform x into y is three, thus, d(x,y) = 3.

We can define a weighted distance that would reflect the difference of the same type of error made on different terminals. Let the weights associated with error transformations on terminal a in a string $\omega_1 a \omega_2$ be defined as follows:[35]

(1) $\omega_1 a \omega_2$ $\left| \dfrac{T_{\sigma,\sigma}(a,b)}{} \right.$ $\omega_1 b \omega_2$, where σ (a,b) is the cost of substituting a for b. Let σ (a,a) $= 0$.

where σ (a,b) is the cost of substituting a for b. Let σ (a,a) $= O$.

(2) $\omega_1 a \omega_2$ $\left| \dfrac{T_{\epsilon,\epsilon}(a)}{} \right.$ $\omega_1 \omega_2$, where ϵ (a) is the cost of deleting a.

where ϵ(a) is the cost of deleting a.

(3) $\omega_1 a \omega_2$ $\left| \dfrac{T_{\phi,\phi}(a,b)}{} \right.$ $\omega_1 b a \omega_2$, where ϕ (a,b) is the cost of inserting b in front of a.

where ϕ (a,b) is the cost of inserting b in front of a.

(4) x $\left| \dfrac{T_{\phi,\phi'}(b)}{} \right.$ x b, where ϕ' (b) is the cost of inserting b at the end of a string x.

where ϕ' (b) is the cost of inserting b at the end of a string x.

Let J be a sequence of transformations used to derive y from x and $|J|$ be defined as the sum of the weights associated with transformations in J. The weighted distance between x and y, $d_w(x,y)$, is defined as

$$d_w (x,y) = \min_{J} \{|J|\} \qquad\qquad (1)$$

An algorithm computing the distance between two strings based on Definition 2 has been proposed by Wagner and Fisher.[56] The algorithm employs dynamic programming techniques. The time and space complexities are both in the order of 0(nm), where n and m are the length of the two strings. The algorithm can easily be modified to compute the weighted distance.[35] Using the distance or the weighted distance between two strings as a similarity measure, recognition of syntactic patterns can be made on the basis of the minimum-distance or the nearest neighbor criterion.[34,35]

Fu and Lu[34,35] have recently suggested using the distance or the weighted distance between two sentences (two strings or two trees*) and between a sentence and a language as a similarity measure between two syntactic patterns and between a syntactic pattern and a class of syntactic patterns. With such similarity measures, cluster analysis can be performed on syntactic patterns using any existing cluster-seeking algorithm.[2,36] As in the decision-theoretic approach, the clustering procedure is nonsupervised. After the cluster analysis is completed, a grammatical inference procedure can be applied to infer a grammar for each cluster, a conventional (nonerror-correcting) parser can be easily constructed for recognition. Consequently, the recognition efficiency is also improved.

When the distance (or weighted distance) is computed on a sentence-to-sentence basis, the use of an error-correcting parser can be avoided in the cluster analysis.[35] Such a direct measure of similarity between two syntactic patterns will result in significant reduction of computations in cluster analysis, although it is often not as flexible as

* Distance on trees has been studied.[58]

that using error-correcting parsers. Also, when the cluster analysis is performed on a sentence-to-sentence basis, the clustering result is usually very sensitive with respect to the variations of pattern size and orientation. Normalizations with respect to pattern size and orientation are required in order to obtain reliable clustering result.

IV. SYNTACTIC RECOGNITION AND ERROR-CORRECTING PARSING

Conceptually, the simplest form of recognition is probably "template-matching". The sentence describing an input pattern is matched against sentences representing each prototype or reference pattern. Based on a selected "matching" or "similarity" criterion, the input pattern is classified in the same class as the prototype pattern that is the "best" to match the input. The structural information is not recovered. If a complete pattern description is required for recognition, a parsing or syntax analysis is necessary. In between the two extreme situations, there are a number of intermediate approaches. For example, a series of tests can be designed to test the occurrence or nonoccurrence of certain subpatterns (or primitives) or certain combinations of them. The result of the tests, through a table lookup, a decision tree, or a logical operation, is used for a classification decision. Recently, the use of discriminant grammars has been proposed for the classification of syntactic patterns.[20]

A parsing procedure for recognition is, in general, nondeterministic and, hence, is regarded as computationally inefficient. Efficient parsing could be achieved by using special classes of languages such as finite state and deterministic languages for pattern description. The tradeoff here between the descriptive power of the pattern grammar and its parsing efficiency is very much like that between the feature space selected and the classifier's discrimination power in a decision-theoretic recognition system. Special parsers using sequential procedures or other heuristic means for efficiency improvement in syntactic pattern recognition have recently been constructed.[21-24]

In practical applications, pattern distortion and measurement noise often exist. Pattern segmentation errors and misrecognitions of primitives (and relations) and/or subpatterns will lead to erroneous or noisy sentences rejected by the grammar characterizing its class. Recently, the use of an error-correcting parser as a recognizer of noisy and distorted patterns has been proposed.[24,28-31] When using an error-correcting parser as a recognizer, the pattern grammar is first expanded to include all the possible errors into its productions. The original grammar is transformed into a covering grammar that generates not only the correct sentences, but also all the possible erroneous sentences. For string grammars, three types of error — substitution, deletion, and insertion — are considered. Misrecognition of primitives (and relations) are regarded as substitution errors, and segmentation errors as deletion and insertion errors. The parser itself is an ordinary parser with a provision added to count the number of error production rules used in parsing a given sentence. This number will be the distance between the input and its nearest neighbor in the language generated by the original grammar. We shall describe some error-correcting parsing algorithms for the following types of languages: regular languages, context-free languages, and tree languages.

A. Error-Correcting Parsing for Regular Languages

Wagner[55] has proposed an error-correcting parser for regular languages that requires time linear to the input string length.[55] A regular language is characterized by the fact that its sentences are precisely the set of sentences acceptable to some finite-state automaton (FSA). Suppose the FSA has scanned the first j symbols of an input string. Then, the only information retained about the symbols already scanned is contained in the "state" of the automaton. The error-correcting algorithm for regular language has taken the advantage of these properties.

Let the input string be $y = a_1 a_2 ... a_m$ and the language under consideration be characterized by a FSA denoted by A. We shall define E(j,Q) to be the minimum number of error transformations needed to change substring $a_1 a_2 ... a_j$ into some substring α. The process will cause A to enter state Q after reading α. E(j,Q) can be computed by given E(j−1, R), a_j, and A:

$$E(j,Q) = \min_{R} \{E(j-1, R) + V(R, Q, a_j)\} \text{ for } j \geq 1$$

where $V(R, Q, a_j)$ is equal to the smallest number of error-transformations which will change the single symbol a_j into a substring that can force A from state R to state Q. The initial condition is

$$E(0,Q) = \begin{array}{ll} 0 & \text{if S is the start state of A} \\ \infty & \text{otherwise} \end{array}$$

The computation proceeds until E(m, Q) is available for all states in A, then the number

$$E(m, T) = \min_{Q \in F} E(m, Q)$$

gives the distance from y to the nearest string acceptable to A where F is the set of final states of A.

The numbers V(R, Q, c) depend only on the FSA. They shall be computed and stored before using the algorithm for recognizing noisy inputs. The storage of V(R, Q, c) values is required for each symbol c in the terminal set and each ordered pair of states R and Q of the FSA.

B. Error-Correcting Parsing for Context-Free Languages

The approach of constructing covering grammar by incorporating error productions into the original grammar is briefly described in this section.[30]

Algorithm 1 — The construction of covering grammar. Input: A context-free grammar $G = (V_N, V_T, P, S)$. Output: A context-free grammar $G' = (V_N', V_T', P', S')$, where is a set of weighted productions.

Methods:

Step 1. $V_N' = V_N \cup \{S'\} \cup \{E_a | a \in V_T\}$, $V_T' \quad V_T$

Step 2. If $A \rightarrow \alpha_o b_1 \alpha_1 b_2 ... b_m \alpha_m$, $m \geq 0$ is a rule in P such that $\alpha_i \in V_N^*$ and $b_i \in V_T$, then add $A \rightarrow \alpha_o E_{b_1} \alpha_1 E_{b_2} ... E_{b_m} \alpha_m$, 0, to P', where each E_{b_i} is a new nonterminal, $E_{b_i} \in V_N'$, and 0 is the weight associated with this production.

  (a) $S' \rightarrow S, 0$

  (b) $S' \rightarrow Sa, \phi'(a)$, for all $a \in V_T'$

  (c) $E_a \rightarrow a, 0$, for all $a \in V_T$

  (d) $E_a \rightarrow b, \sigma(a,b)$, for all $a \in V_T'$, $b \in V_T'$ and

  (e) $E_a \rightarrow \lambda, \varepsilon(a)$, for all $a \in V_T$

  (f) $E_a \rightarrow b E_a, \phi(a,b)$, for all $a \in V_T$, $b \in V_T'$, where $\sigma, \varepsilon, \phi$ and ϕ' are weights associated with error transformations as defined in Section III.

In Algorithm 1 the production rules added in Step 3(b), 3(d), 3(e), and 3(f) are called error productions. Each error production corresponds to one type of error transformation on a symbol in V_T. Therefore the distance calculated in terms of error transformations can be measured by counting the number of error productions used in a derivation.

If the pattern grammar is nonstochastic, the minimum-distance (or minimum weighted distance) criterion for error-correcting parsing can be applied. The parser described[30] is a modified Earley parser with a provision added to count the number of error productions used (or to accumulate the weights). The parsing algorithm always chooses the derivation that associated with the least number of error productions used (or the least weight). On the other hand, if the grammar is stochastic, the maximum-likelihood and Bayes criteria can be used.[24,30] One difficulty about this approach is the parsing time, in particular, when all the three types of error are considered.

The sequential parsing procedure suggested by Persoon and Fu[23] has been applied to error-correcting parser to reduce the parsing time.[24] By sacrificing a small amount of error-correcting power (that is, allowing a small error in parsing), a parsing could be terminated much earlier before a complete sentence is scanned. The tradeoff between the parsing time and the error committed can be easily demonstrated. In addition, error-correcting parsing for transition network grammars[30] and tree grammars[31] has also been studied. For tree grammars, five types of error — substitution, deletion, stretch, branch, and split —are considered. The original tree pattern grammar is expanded by including the five types of error transformation rule. The tree automaton constructed according to the expanded tree grammar and the minimum-distance criterion is called an error-correcting tree automaton (ECTA). When only substitution errors are considered, the structure of the tree to be analyzed remains unchanged. Such an error-correcting tree automaton is called a "structure-preserved error-correcting tree automaton" (SPECTA).[32,54] Another approach to reduce the parsing time is the use of parallel processing.[33]

V. SYNTACTIC APPROACH TO SHAPE ANALYSIS

Recently, syntactic methods have been applied to both shape description and recognition. Pavlidis and Ali[37] have proposed a general model of syntactic shape analyzer. The first major component of the model is a curve-fitting algorithm that achieves the noise elimination and data reduction. The split-and-merge algorithm is used to obtain a polygonal approximation of the boundary of the original picture or object. It is assumed that the boundaries of the objects of interest consist of concatenations of the following subpatterns or nonterminals: QUAD (arcs approximated by a quadric curve), TRUS (sharp protrusions or intrusions), LINE (long line segments), and BREAK (short segments with no regular shape). Each of the nonterminals has a set of attributes as its semantic information. The production rules of the proposed general shape grammar consist of both syntactic and semantic rules. Stochastic finite automata are used as parsers for shape recognition.

Another method recently proposed for syntactic shape description and recognition is the use of attributed grammars.[18] Two types of primitive with attributes are proposed. The first type is a curve segment with its direction (the vector from the starting point to the end point), total length, total angular change, and a measure of its symmetry as the four attributes. The second type of primitive is an angle primitive with its attribute specified by the angular change at the concatenating point of two consecutive curve segments. Finite-state and context-free attributed grammars are used for shape description and recognition. Each production rule of the attributed grammar has a symbolic part like the conventional grammar rule and a semantic part for processing the attributes of the terminals and nonterminals in the symbolic part. The primitive extraction process is embedded in the parsing of the strings describing the boundaries of objects. Modified Earley parser and finite automata are used as shape recognizers.

VI. APPLICATIONS TO WAVEFORM AND SIGNAL PROCESSING

Syntactic pattern recognition has been applied to waveform analysis, ECG interpretation, speech recognition and understanding,* character recognition, fingerprint classification, recognition of two-dimensional mathematical notation, modeling of Earth Resources Satellite data, machine parts recognition, and automatic visual inspection.[6,8,32,37-44] In this paper, we briefly review some of the recent applications of syntactic methods to waveform and signal processing.

Waveforms are basically one-dimensional signals, which appear to be naturally suitable for the application of syntactic methods. A waveform could be represented by a concatenation of waveform segments. However, the selection of primitives (basic waveform segments) and subpatterns could be quite different from different application points of view. Linear and/or quadric segments through functional approximation have been proposed as waveform primitives.[6] Ehrich and Foith[45] have proposed the use of a relational tree to describe a waveform in terms of its peaks and valleys. Sankar and Rosenfeld[46] have recently proposed an alternative method of using a peak relational tree in terms of fuzzy connectivity to describe waveforms.

In Fu,[6] Horowitz has proposed a deterministic context-free grammar that can be used to recognize positive and negative peaks in a waveform represented by a string of "positive slope", "negative slope", and "zero slope" primitives. The grammar is

$$G = (V_N, V_T, P, W)$$

where $V_N = \{W, <W>, <P^+>, <P^->, <p_1>, <n_1>,$
$\qquad\qquad <p_2>, <n_2>, <z>\}$

$\qquad V_T = \{p/, n\backslash, \underline{O}\}$

and P: $W \rightarrow <z><w><z>$

$\qquad W \rightarrow <z><w>$

$\qquad W \rightarrow <w><z>$

$\qquad W \rightarrow <w>$

$\qquad W \rightarrow <z>$

$\qquad <w> \rightarrow <P^+>$

$\qquad <w> \rightarrow <P^->$

$\qquad <w> \rightarrow <p_1>$

$\qquad <w> \rightarrow <n_1>$

$\qquad <P^+> \rightarrow ><P^-><z><n_1>$

$\qquad <P^+> \rightarrow <P^-><n_1>$

$\qquad <P^+> \rightarrow <p_1><z><n_1>$

$\qquad <P^+> \rightarrow <p_1><n_1>$

$\qquad <P^-> \rightarrow <P^+><z><p_1>$

$\qquad <P^-> \rightarrow <P^+><p_2>$

$\qquad <P^-> \rightarrow <n_1><z><p_1>$

$\qquad <P^-> \rightarrow <n_1><p_1>$

$<p_1> \rightarrow <p_1> p$

$<p_1> \rightarrow <p_2> p$

$<p_1> \rightarrow p$

$<n_1> \rightarrow <n_1> n$

$<n_1> \rightarrow <n_2> n$

$<n_1> \rightarrow n$

$<p_2> \rightarrow <p_1> O$

$<p_2> \rightarrow <p_2> O$

$<n_2> \rightarrow <n_1> O$

$<n_2> \rightarrow <n_2> O$

$<z> \rightarrow <z> O$

$<z> \rightarrow O$

---

* For a detailed review of the recent progress in speech recognition, and ECG processing, refer to References 41 and 53.

When parsing a specific string describing a waveform, a positive peak is recongized if and only if a section of the string is completely reduced by some production to the nonterminal <P⁺>. Similarly, a negative peak is recognized if and only if a section of the input string is reduced to the nonterminal <p⁻>.

Stockman et al.[22] have suggested the use of a waveform parsing system to analyze carotid pulse waves. A typical carotid pulse wave and its hierarchical structure representation are shown in Figure 4A and 4B, respectively. The context-free grammar generating various types (structures) of carotid pulse waves is given below.

$$G = (V_N, V_T, P, <\text{CAROTID PULSE WAVE}>)$$

where V_N = {<CAROTID PULSE WAVE>, <SYSTOLIC>,
          <DIASTOLIC>, <MAXIMA>, <M1>, <M1>, <M2>,
          <M3>, <DICROTIC WAVE>, <POS WAVE>,
          <NEG WAVE>}

V_T = {UPSLOPE, LARGE-POS, LARGE-NEG,
      MED-POS, MED-NEG, TRAILING-EDGE
      HOR, CAP, PEK, CCUP, VCUP
      RSHOLD, LSHOLD}

Primitives:

UPSLOPE_____Line: long, large positive slope
LARGE-POS_____Line: medium length, large positive slope
LARGE-NEG_____Line: medium length, large negative slope
MED-POS_____Line: medium length and positive slope
MED-NEG_____Line: medium length and negative slope
TRAILING-EDGE__Line: long, medium negative slope
HOR_____Line: short, near 0 slope
CAP_____Parabola
PEK_____Parabola
CCUP_____Parabola
VCUP_____Parabola
RSHOLD_____Parabola: right half of parabolic maxima
LSHOLD_____Parabola: left half of parabolic maxima

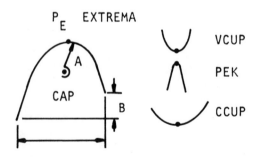

FIGURE A.   Generic cap morph; (B) variations.

and P:

<CAROTID PULSE WAVE> → <SYSTOLIC><DIASTOLIC>

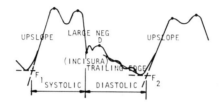

UPSLOPE CAP CCUP CAP LARGE-NEG PEK CAP TRAILING-EDGE

A

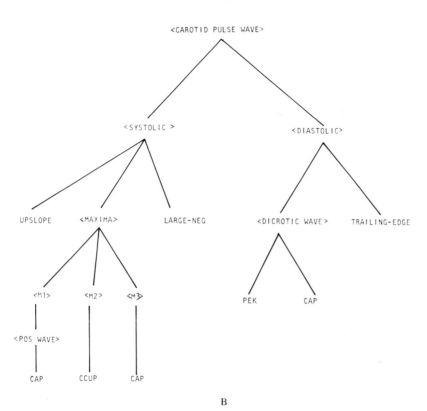

B

FIGURE 4.   A typical carotid wave and its hierarchical structure representation. (A) Up-slope cap ccup cap large-neg pek cap trailing edge; (B) structural represenation of a carotid pulse wave.

<SYSTOLIC> → UPSLOPE <MAXIMA> LARGE-NEG
<MAXIMA> → <M1><M2><M3>
<MAXIMA> → MED-POS <M3>
<MAXIMA> → <M1> MED-NEG
<DIASTOLIC> → TRAILING-EDGE
<DIASTOLIC> → <DICROTIC WAVE> TRAILING-EDGE
<DICROTIC WAVE> → CAP, <DICROTIC WAVE> → HOR
<DICROTIC WAVE> → PEK, <DI-ROTIC WAVE> → PEK CAP
<M1> → LSHOLD, <M1> → <POS WAVE>
<M2> → CCUP, <M2> → <NEG WAVE>
<M3> → RSHOLD, <M3> → CAP
<POS WAVE> → CAP, <POS WAVE> → CAP LARGE-NEG
<NEG WAVE> → VCUP, <NEG WAVE> → VCUP LARGE-POS

Le Chevalier et al.[47] have recently proposed the use of syntactic decoding method (error-correcting string parser with substitution error only) for syntactic signal processing. Waveform segments (e.g., segments of sinusoids) are selected as primitives. A signal is described as a sequence (concatenation) of primitives. In addition to the distance suggested in error-correcting parsing, the correction is also used as a similarity measure between two strings. Practical applications include signal detection, radar target identification and adaptive antenna processing.

VII. CONCLUDING REMARKS

We have briefly reviewed some recent advances in the area of syntactic pattern recognition and its application to signal processing. Due to noise and distortions in real world patterns, syntactic approach to pattern recognition was regarded earlier as only effective in handling abstract and artificial patterns. However, with the recent development of distance or similarity measures between syntactic patterns and error-correcting parsing procedures, the flexibility of syntactic methods has been greatly expanded. Errors occurring at the lower-level processing of a pattern (segmentation and primitive recognition) could be compensated at the higher level using structural information. Using a distance or similarity measure, nearest-neighbor and k-nearest-neighbor classification rules can be easily applied to syntactic patterns. Furthermore, with a distance or similarity measure, a clustering procedure can be applied to syntactic patterns. Such a nonsupervised learning procedure can also be very useful for grammatical inference in syntactic pattern recognition.[19,34]

It has been noticed from the recent advances that semantic information has been used more and more with the syntax rules in characterizing patterns. Quite often, semantic information involving spatial information can be expressed syntactically such as attributed grammars and relational trees and graphs.[3,10,13,18,43,45] Parsing efficiency has become a concern in syntactic recognition. Special grammars and parallel parsing algorithms have been suggested for speeding up the parsing time. Structural information of an image can also be used as a guide in the segmentation process through the syntactic approach.[40,48] Syntactic representation of patterns such as hierarchical trees and relational graphs should also be very useful for database organization. Several recent publications have already shown such a trend.[49-52]

In some applications, both the decision-theoretic and the syntactic methods may be used. One possibility is to use a decision-theoretic method for the recognition of primitives and then to use a syntactic method for the recognition of subpattern and pattern itself. The second possibility is the use of stochastic grammars with which the syntactic recognition is made in the decision-theoretic sense (maximum-likelihood or Bayes). The term "mixed or combined approach" is often used to denote such an approach.

REFERENCES

1. **Fu, K. S. and Rosenfeld, A.,** Pattern Recognition and Image Processing, *IEEE Trans. Comput.,* C-25(12), 1336, 1976.
2. **Fu, K. S.,** *Digital Pattern Recognition,* Springer-Verlag, Basel, 1976.
3. **Fu, K. S.,** *Syntactic Methods in Pattern Recognition,* Academic Press, New York, 1974.
4. **Hanakata, K.,** Feature selection and extraction for decision theoretic approach and structural approach in *Pattern Recognition-Theory and Application,* Fu, K. S. and Whinston, A. B., Eds., Noordhoff International, Leyden, The Netherlands, 1977, 133.

5. Chen, C. H., On statistical and structural feature extraction, in *Pattern Recognition and Artificial Intelligence,* Chen, C. H., Ed., Academic Press, New York, 1976, 135.

6. Fu, K. S., *Syntactic Pattern Recognition Applications,* Springer-Verlag, Basel, 1977.

7. Gips, J., *Shape Grammars and Their Use,* Birkhauser Verlag, Basel, 1975.

8. Fu, K. S., Tree languages and syntactic pattern recognition, in *Pattern Recognition and Artificial Intelligence,* Chen, C. H., Ed., Academic Press, New York, 1976.

9. Fu, K. S. and Booth, T. L., Grammatical inference-introduction and survey, *IEEE Trans. Syst. Man Cybern.,* Part 1, SMC-5, 59, 1975; Part II, SMC-5, 409, 1975.

10. Chou, S. M. and Fu, K. S., Inference for transition network grammars, in *Proc. 3rd Int. Jt. Conf. on Pattern Recognition,* IEEE, Piscataway, N.J., 1976.

11. Porter, G. B., Grammatical inference based on pattern recognition, in *Proc. 3rd Int. Jt. Conf. on Pattern Recognition,* IEEE, Piscataway, N.J., 1976.

12. Miclet, L., Inference of regular expressions, *Proc. 3rd Int. Jt. Conf. on Pattern Recognition,* IEEE, Piscataway, N.J., 1976.

13. Brayer, J. M. and Fu, K. S., Some multidimensional grammar inference methods, in *Pattern Recognition and Artificial Intelligence,* Chen, C. H., Ed., Academic Press, New York, 1976, 29.

14. Brayer, J. M. and Fu, K. S., A Note on the k-tail method of tree grammar inference, *IEEE Trans. Syst. Man Cybern.,* SMC-7(4), 293, 1977.

15. Barrero, A. and Gonzalez, R. C., A tree traversal algorithm for the inference of tree grammars, in *Proc. 1977 IEEE Comput. Soc. Conf. on Pattern Recognition and Image Processing,* IEEE, Piscataway, N.J., 1977.

16. Lee, H. C. and Fu, K. S., A Syntactic Pattern Recognition System with Learning Capability, in Proc. 4th Int. Symp. on Comput. and Information Sciences (COINS-72), Bal Harbour, Fla., December 14 to 16, 1972.

17. Keng, J. and Fu, K. S., A System of Computerized Automatic Pattern Recognition for Remote Sensing, in Proc. 1977 Int. Comput. Symp., Taipei, Taiwan, December 27 to 29, 1977.

18. You, K. C. and Fu, K. S., Syntactic Shape Recognition Using Attributed Grammars, in Proc. 8th EIA Symp. on Automatic Imagery Pattern Recognition, Gaithersburg, Md., April 3 to 4, 1978.

19. Lu, S. Y. and Fu, K. S., Stochastic tree grammar inference for texture synthesis and discrimination, *Comput. Graphics Image Process.,* 9, 234, 1979.

20. Page, C. and Filipski, A., Discriminant grammars, an alternative to parsing for pattern classification, in *Proc. 1977 IEEE Workshop on Picture Data Description and Management,* IEEE, Piscataway, N.J., 1977.

21. Pavlidis, T., Syntactic feature extraction for shape recognition, in *Proc. 3rd Int. Jt. Conf. on Pattern Recognition,* IEEE, Piscataway, N.J., 1976.

22. Stockman, G., Kanel, L. N., and Kyle, M. C., Structural pattern recognition of carotid pulse waves using a general waveform parsing system, *Commun. ACM,* 19(12), 688, 1976.

23. Persoon, E. and Fu, K. S., Sequential Classification of Strings Generated by SCFG's, *Int. J. Comput. Inf. Sci.,* 4, 205, 1975.

24. Lu, S. Y. and Fu, K. S., Stochastic error-correcting syntax analysis for recognition of noisy patterns, *IEEE Trans. Comput.,* C-26(12), 1268, 1977.

25. Pavlidis, T., Syntactic pattern recognition on the basis of functional approximation, in *Pattern Recognition and Artificial Intelligence,* Chen, C. H., Ed., Academic Press, New York, 1976, 389.

26. Joshi, A. K., Remarks on some aspects of language structure and their relevance to pattern analysis, *Pattern Recognition,* 5(4), 365, 1973.

27. Bhargava, B. K. and Fu, K. S., Transformation and Inference of Tree Grammars for Syntactic Pattern Recognition, in Proc. 1974 IEEE Int. Conf. on Cybern. and Soc., Dallas, October 1974.

28. Fung, L. W. and Fu, K. S., Stochastic syntactic decoding for pattern classification, *IEEE Trans. Comput.,* C-24, 662, 1975.

29. Thomason, M. G. and Gonzalez, R. C., Error detection and classification in syntactic pattern structures, *IEEE Trans. Comput.,* C-24, 93, 1975.

30. Fu, K. S., Error-correcting parsing for syntactic pattern recognition, in *Data Structure, Computer Graphics and Pattern Recognition,* Klinger, A., Fu, K. S., and Kunii, T., Eds., Academic Press, New York, 1977.

31. Lu, S. Y. and Fu, K. S., Error-correcting tree automata for syntactic pattern recognition, in *Proc. 1977 IEEE Conf. on Pattern Recognition and Image Processing,* IEEE, Piscataway, N.J., 1977.

32. Lu, S. Y. and Fu, K. S., Structure-preserved error-correcting tree automata for syntactic pattern recognition, in *Proc. 1976 IEEE Conf. on Decision and Control,* IEEE, Piscataway, N.J., 1976.

33. Chang, N. S. and Fu, K. S., Parallel Parsing of Tree Languages, in *Proc. 1978 IEEE Comput. Soc. Conf. on Pattern Recognition and Image Processing,* IEEE, Piscataway, N.J., 1978.

34. Fu, K. S. and Lu, S. Y., A clustering procedure for syntactic patterns, *IEEE Trans. Syst. Man Cybern.,* SMC-7(10), 734, 1977.

35. Lu, S. Y. and Fu, K. S., A sentence-to-sentence clustering procedure for pattern analysis, *IEEE Trans. Syst. Man Cybern.*, SMC-8(5), 381, 1978.
36. Duda, R. O. and Hart, P. E., *Pattern Classification and Scene Analysis*, John Wiley & Sons, New York, 1972.
37. Ali, F. and Pavlidis, T., Syntactic recognition of handwritten numerals, *IEEE Trans. Syst. Man Cybern.*, SMC-7(7), 537, 1977.
38. Brayer, J. M. and Fu, K. S., Application of a web grammar model to an ERTS picture, in *Proc. 3rd Int. Jt. Conf. on Pattern Recognition*, IEEE, Piscataway, N.J., 1976.
39. Li, R. Y. and Fu, K. S., Tree System Approach to LANDSAT Data Interpretation, in Proc. Symp. on Machine Process. of Remotely Sensed Data, West Lafayette, Ind., June 29 to July 1, 1976.
40. Keng, J. and Fu, K. S., A Syntax-Directed Method for Land-Use Classification of LANDSAT Images, in Proc. Symp. on Current Mathematical Problems in Image Science, Monterey, Calif., November 10 to 12, 1976.
41. DeMori, R., On speech recognition and understanding in *Pattern Recognition: Theory and Application*, Fu, K. S. and Whinston, A. B., Eds., Noordhoff, Leyden, Netherlands, 1977, 289.
42. Jakubowski, R. and Kasprzak, A., A syntactic description and recognition of rotary machine elements, *IEEE Trans. Comput.*, C-26(10), 1039, 1977.
43. Jarvis, J. F., Regular expressions as a feature selection language for pattern recognition, in *Proc. 3rd Int. Jt. Conf. on Pattern Recognition*, IEEE, Piscataway, N.J., 1976.
44. Mundy, J. L. and Joynson, R. E., Automatic visual inspection using syntactic analysis, in *Proc. 1977 IEEE Comput. Soc. Conf. on Pattern Recognition and Image Processing*, IEEE, Piscataway, N.J., 1977.
45. Ehrich, R. W. and Foith, J. P., Representation of random waveforms by relational trees, *IEEE Trans. Comput.*, C-25, 725, 1976.
46. Sankar, P. V. and Rosenfeld, A., Hierarchical Representation of Waveforms, TR-615, Computer Science Center, University of Maryland, College Park, 1977.
47. Le Chevalier, F., Bobillot, G., and Fugier-Garrel, C., Syntactic Signal Processing, in 1978 Int. Symp. on Inf. Theory, Ithaca, N.Y., October 10 to 14, 1978.
48. Tsuji, S. and Fujiwana, R., Linguistic Segmentation of Scenes into Regions, in Proc. 2nd Int. Jt. Conf. on Pattern Recognition, Copenhagen, August 13 to 15, 1974.
49. Kunii, T., Weyle, S., and Tenenbaum, J. M., A Relational Data Base Scheme for Describing Complex Pictures with Color and Texture, in Proc. 2nd Int. Jt. Conf. on Pattern Recognition, Copenhagen, August 13 to 15, 1974.
50. Bonczek, R. H. and Whinston, A. B., Picture processing and automatic data base design, *Comput. Graphics Image Process.*, 5(4), 1976.
51. Kashyap, R. L., Pattern recognition and data base, in *Proc. 1977 IEEE Comput. Soc. Conf. on Pattern Recognition and Image Processing*, IEEE, Piscataway, N.J., 1977.
52. Chang, S. K., Syntactic description of pictures for efficient storage retrieval in a pictorial data base, in *Proc. 1977 IEEE Comput. Soc. Conf. on Pattern Recognition and Image Processing*, IEEE, Piscataway, N.J., 1977.
53. Beforte, G., DeMori, R., and Ferraris, F., A contribution to the automatic processing of electrocardiogram using syntactic methods, *IEEE Trans. Biomed. Eng.*, BME-26, 1979.
54. Lu, S. Y. and Fu, K. S., A syntactic approach to texture analysis, *Computer Graphics Image Process.*, 7(3), 303, 1978.
55. Wagner, R. A., Order-n correction for regular languages, *Commun. ACM*, 17(5), 1974.
56. Wagner, R. A. and Fisher, M. J., The string to string correction problem, *J. ACM*, 21(1), 168, 1974.
57. Fu, K. S. and Lu, S. Y., Size normalization and pattern orientation problem in syntactic clustering, *IEEE Trans. Syst. Man, Cybern.*, SMC-9, 55, 1979.
58. Lu, S. Y., A Tree-to-tree distance and its application to cluster analysis, *IEEE Trans. Pattern Anal. Mach. Intelligence*, 1(2), 219, 1979.

Chapter 6

SPEECH PROCESSING

Alistair D. C. Holden

TABLE OF CONTENTS

I. INTRODUCTION

Digital processing of speech data involves many levels of abstraction. The digitized speech signal first undergoes signal processing using techniques such as zero-crossing analysis, filter-bank spectral analysis, fast Fourier transform (FFT) analysis, linear prediction coefficient (LPC) analysis, or homomorphic filtering. This level produces a set of features that encodes the information needed for recogniton of the phones in an utterance.

Typical of the features used are the vocal tract resonant frequencies (formants) and their bandwidths, voiced or unvoiced decisions, relative energy levels, fundamental frequency (excitation frequency or "pitch" of the vocal chord impulses), etc. For speech recognition, formants of higher frequency than the first three are unnecessary. Occasionally, features are used that are more directly related to the signal processing method. Thus, the set of coefficients from an LPC analysis, an orthogonalized set derived from them, or the complex cepstrum may be used. Also, to reduce the dimensionality of the decision problem, a threshold may be used to reduce each feature to a binary variable.

For a given utterance, a feature set may be generated every centisecond or so; the next step is to generate a string of phonemes (basic linguistic units of speech) derived from the relationships between features. The reliability at this stage is usually quite low so that, instead of generating one linear string of phonemes, a string with several alternative phonemes at each step is generated. The process of choosing one or more particular strings of phonemes to form words and sentences is guided by the syntax of the language involved. The syntax specifies which sequences of phonemes can occur to form legal words and sentences and greatly reduces the number of sequences to be tried. Clearly, the constraints imposed by the syntax lead to greater reliability in the recognition process. Although there is no usable syntax for all of the English language, subsets of English can be described in this way, and many voice recognition tasks allow the designer to specify both the vocabulary and the syntax of the language to be used.

Progress has been made over the last 10 years in using the meaning (semantics) of an utterance to guide natural language (and speech) recognition by computer. The term "speech understanding" is used when semantics are employed, and this is a very active area of artificial intelligence research. A simple example of the use of meaning to help in speech recognition is in the examination of the set of legal chess moves in a given chess game in the HEARSAY I system[1] to decide between conflicting interpretations of an utterance given to specify the next move. The effect of this was to throw out many of the wrong candidate sentences, but once in a while it actually led to the wrong move.

There are similarities between the speech recognition and image analysis areas. However, in spite of the saying that "a picture is worth a thousand words," speech understanding probably reaches higher levels of abstraction since, ultimately, a symbolic representation of any concept of human thought is involved.

II. MODEL OF SPEECH PRODUCTION

There is, fortunately, an accurate model of the human speech production system. The accuracy of the model can be verified by using it to synthesize the numerous classes of speech sounds (phones).

In speech, when "voiced" sounds (such as vowels) are produced, the throat, mouth, and nasal cavities are excited by a series of air pressure impulses caused by the opening and closing of the vocal cords. The excitation function can thus be represented by a train of delta functions. When "unvoiced" sounds are produced, the excitation is equivalent to white noise. Most of the sounds of speech can be modeled quite well by the response of three loosely coupled resonant cavities to either a pseudoperiodic series of impulses or to noise.

The intelligence in speech comes from the relatively slow variations in the frequencies of resonance (i.e., "formants", usually labeled f_1, f_2, f_3) and the associated bandwidths of the cavities and the fundamental or "pitch" frequency (f_o). If the values of the first three formants and their bandwidths and the pitch frequency are specified every 5 msec or so, intelligible speech can be synthesized. The model of speech production is thus a system with three or so resonances and slowly time-varying parameters that is excited either by a series of impulses (whose period varies as the pitch-period changes) or by noise.

The source model described above is the basis for methods that can substantially reduce the bandwidth needed for speech communication. Instead of sending the analog signal corresponding to the speech, with all of the implied redundancy, it is sufficient to send the encoded values of the parameters from the speech production model. Thus, if the voiced/unvoiced state, the intensity, pitch period, and resonance parameters are sent every 5 msec, good quality speech can be reconstituted at the receiver. Using direct A/D conversion and transmitting a digital version of the analog signal (i.e., PCM), a rate of 64 kilobit/sec is needed. With "speech compression" where only the source model parameters are transmitted, intelligible speech can be obtained with the greatly reduced rate of 1 kilobit/sec. It is peculiar that the idea of speech compression using a source model did not occur many years ago. The method of sending a replica of the incoming signal over a channel is very wasteful of bandwidth. The situation is similar to the case where it is known that a 10 MHz signal is needed at the receiver at various times. Instead of sending the 10 MHz signal over the channel whenever it is needed, it is sufficient to send a one-bit signal whenever the signal is required. This can be used to switch a 10 MHz generator at the receiving end "on" or "off". There is no need to repeatedly send information over a channel when enough is known of the source to reconstruct the signal with much less information. This is true for speech, video, or any other kind of data. Of course, there is a penalty in the increased cost of hardware required for the encoding and decoding process. However, the cost of such hardware will soon be low.

III. ANALYTICAL METHODS

For voiced speech, it is clear that if the pitch period can be found, then all that is required for speech recognition is the spectrum of the signal within each pitch period. The signal within each pitch period is, in fact, the impulse response of the system at that time. The spectrum reveals the resonances. This would be a "pitch synchronous" method. However, the first problem here is that the detection of the beginning and ending of each pitch period is not too easy.

A. The Cepstrum

The cepstrum of the signal (discussed in Chapter 7) can be used for pitch detection, but this method is computationally expensive. The real part of the complex cepstrum of a short-time segment of voiced speech (about 30 ms) usually has a peak in the high-order range, which indicates the pitch period. Also, the imaginary part of the complex spectrum usually has a "doublet" at the same point.

Incidentally, the cepstrum representation of speech is useful also for speech recognition[2] and for speaker identification. For speech recognition, the voiced/unvoiced situation can be determined from the high orders, and the particular phoneme spoken can be recognized from the low orders. The low order cepstrum represents the smoothed spectrum and is quite invariant for the same vowel from the same speaker. Separating the low and high orders of the cepstrum corresponds to splitting the excitation representation from the response representation. For voiced sounds, the excitation spectrum consists of a line spectrum with intensity that decreases with frequency. The fundamental frequency is also much lower (about 500 Hz) than the resonances of the impulse responses. To extract the resonances (formants) for speech recognition purposes, the multiplicative contribution of the excitation spectrum (which causes many periodic local peaks) has to be removed. This can readily be done by computing the cepstrum, subtracting the peak in the high orders due to the excitation function (i.e., high-pass "liftering"), and taking the Fourier transform of the result. The end product is close to the smoothed spectrum of the impulse response. (The actual-response spectrum would be produced if only a single pitch-period had been taken at the start.) The success of this is due to the "log" operator in the cepstrum, which for a time function x(t) is defined by $\mathcal{F}^{-1}\{\log|\mathcal{F}\{x(t)\}|\}$. If FFT hardware is available, the cepstrum can be computed in real time.

B. Linear Predictive Analysis

The development of linear prediction methods, mainly as a result of the work of Atal and Hanauer[3] and Itakura and Saito,[4,5] has played a very important part in the acceleration of results in speech processing in recent years. Linear prediction coefficient (LPC) analysis leads to efficient methods of computing a "smoothed" short-time spectrum, which allows the formant peaks and (less accurately) their bandwidths to be readily detected. It is also the basis of a good method ("SIFT", described later) for finding the impulse excitation points in voiced speech and, hence, the pitch period.

Although the basic principles of prediction were discussed in Chapter 3, a short treatment will be given here since the application to speech analysis has some unique features.

If the speech signal samples $\{\chi_n\}_{n=0}^{N-1}$ are approximated by $\hat{\chi}_n = \sum_{k=1}^{p} \hat{a}_k \chi_{n-k}$ such that the total squared error

$$e = \sum_{n=p}^{n-1} (x_n - \hat{x}_n)^2$$

is minimized, then it is easy to show that the linear prediction coefficients $\{a_k\}_{k=1}^{p}$ are given by the solution to the following set of p linear equations:

$$\sum_{k=1}^{p} a_k \phi_{ik} = \phi_{i0}, \quad 1 \leqslant i \leqslant p$$

where

$$\phi_{ik} = \sum_{n=p}^{n-1} x_{n-i} \, x_{n-k}, \quad 1 \leqslant i \leqslant p$$
$$0 \leqslant k \leqslant p$$

If the signal, $\{\chi_n\}$ is stationary, then the linear prediction coefficients can be obtained from the autocorrelation function, R_i, of the signal, where

$$R_i = \frac{1}{N} \sum_{n=1}^{N-|i|} x_n \, x_{n+|i|}$$

(assuming that χ_n is zero for n outside the range $1 \leqslant n \leqslant N$) by solving the p linear equations

$$\sum_{k=1}^{p} a_k \, R_{|i-k|} = R_i, \quad 1 \leqslant i \leqslant p$$

This is termed the stationary method as compared with the previous nonstationary approach, which is called the autocovariance method. A good discussion of this is given by Chandra and Lin[6] and an excellent detailed tutorial treatment of this, and speech processing as a whole, is given by Rabiner and Schafer.[7]

A fast recursive computational method for $\{a_k\}$ (Durbin's method) which uses on the order of p^2 operations, is described by Markel and Gray.[8] The method exploits the fact that the autocorrelation matrix (which normally would have to be inverted to solve for $\{a_k\}$ is of Toeplitz form having the same element along all left to right descending diagonals, i.e., $R = [r_{ij}]$, with $r_{i+1,j+1} = r_{ij}$, $0 \leqslant i, j \leqslant (p-1)$, also $r_{oj} = R_j$, $0 \leqslant j \leqslant (p-1)$.

The algorithm (which is repeated for i = 1 to p) is: (Initially, the error E_0 is R_0)

Step 1

$$\alpha_i = \left[R_i - \sum_{j=1}^{i-1} a_j^{(i-1)} R_{|i-j|} \right] / E_{i-1}, \quad 1 \leqslant i \leqslant p$$

Step 2.

$$A_i^{(i)} = \alpha_i$$

Step 3.

$$A_j^{(i)} = A_j^{(i-1)} - \alpha_i \, A_{i-j}^{(i-1)} \quad 1 \leqslant j \leqslant_{i-1}$$

Step 4.

$$E_i = (1-\alpha_i^2) \, E_{i-1}$$

The final solution is given by

$$a_i = a_i^{(p)}, \quad 1 \leqslant i \leqslant p$$

The partial results obtained at any point in this algorithm are valid for a prediction of the order reached. Thus, E_i in Step 4 gives the prediction error for a predictor of the order reached; the $a_j^{(i)}$ term in Step 3 is the jth coefficient for a predictor of order i. These terms are called PARCOR (Partial correlation) values.

C. The SIFT Algorithm

The SIFT (Simple Inverse Filtering) Algorithm developed by Markel[9] is an excellent example of the use of linear prediction to locate the beginning of each pitch period. The following operations are applied in sequence:

Step 1. The speech signal is passed through a 900 Hz low-pass filter. (This includes the usual frequency range of the voicing excitation.)

Step 2. Every fifth sample is selected, and the rest are dropped (i.e., a 5:1 decimation).

Step 3. A four coefficient ($p = 4$) predictor is computed, and the signal is passed through the corresponding inverse filter. (This allows two complex pole pairs and, hence, two resonances and tends to spectrally flatten the signal.)

Step 4. The short-time autocorrelation function of the result is completed.

Step 5. The greatest peak in the autocorrelation function is found. (This will give the pitch period.)

The algorithm works well for low-pitched speakers but not so well for the women's and children's voices, which tend to be high pitched.

IV. PROSODIC FEATURES

The usual synthetic speech from a computer sounds quite robot-like. The main reason for this is the absence of emphasis or linguistic stress. The emphasis perceived in speech is rather a complex phenomenon, because not only does it depend on purely acoustic cues, but also it is related to the meaning and to the expectations of the listener.

However, work with nonsense syllables[10] has shown that linguistic stress, when no meaning is attached to an utterance, is related to syllable duration, intensity, fundamental frequency, f_o, and the first two derivatives of f_o. The degree to which any one of these five factors is used depends upon the individual and the content.[11-14]

A. Linguistic Stress In Automatic Speech Recognition

In speech recognition systems, the use of stressing information is important. Highly stressed syllables can be used as anchor points for syntactic processing, and the ordering of stress can be used to reflect the syntactic relationship of words in the utterance. Highly stressed syllables also have different features from syllables with less stress, so the stress level is an important recognition parameter.

1. Locating Anchor Points

On the phonetic level, the amount of stressing exerted on each syllable by a speaker directly affects the conditions for reliable recognition. For example, it is known that the vowel quality in unstressed syllables often reduces to be schwa-like,[15] causing the vowel to lose its distinctive vowel quality. In extreme conditions of stress reduction, an unstressed vowel can be totally missing in the pronunciation. Stressing information yields a good probabilistic measure, indicating first whether the vowel is present in the acoustic signal or not and second, whether the recognition of the vowel is reliable or not. Hence, information on the magnitudes of syllable stress allows the speech recognition system to benefit from the expected reliability in recognition of highly stressed vowels and to handle the unstressed vowels in any special manner necessary.

In a system using syntactic analysis, specially chosen syllables called "anchor points" must first be secured so that processing may proceed from these syllables on either side of their temporal locations. These anchor points are very important in the analysis because incorrectly recognized anchor points can easily misdirect the interpretation and lead to complete failure in sentence recognition. Usually, the anchor points are chosen as those syllables with the highest recognition score. If the stressing information of the syllables is known, then the highly stressed syllables can readily be used as anchor points for further processing of the rest of the sentence, which consists of less-stressed syllables.

2. Composite Expression For Stress Estimation

The stress magnitude S, on a syllable, can be computed from the acoustic signal. This method was devised after a large amount of data from experiments correlating human judgments with computer estimated values was analyzed.

$$S = (aF + bI_{norm}) \log_{10} D_{norm}$$

where

$$F = (F_M + 0.5F_1 + 0.5F_{LD})_{norm}$$

F_M = mean syllable fundamental frequency
F_1 = fundamental frequency increase due to inflection
F_{LD} = local deviation of fundamental frequency

The two modifying constants, a and b, adjust the relative dominance of fundamental frequency and intensity, allowing for differences between speakers who emphasize tonal variations and those who emphasize intensity variations. Normalized values are defined as follows:

F_{norm} = $(F_m - 0.5F_{av})/F_{av}$
F_m = mean value of fundamental frequency within syllable
F_{av} = mean fundamental frequency over all syllables
I_{norm} = $(I_m - I_{min})/(I_p - I_{min})$
I_m = mean value of intensity within syllable
I_{min} = noise threshhold
I_p = peak intensity value within syllable
D_{norm} = $(10 D_m)/D_{av}$
D_m = mean syllable duration within syllable
D_{av} = mean syllable duration over all syllables

B. Example Of Stress Estimation From Acoustic Parameters

The result of two sets of experiments will be given to illustrate this method.[10] In Experiment I, higher level linguistic influences were removed by using the sequence of nonsense syllables, /ba ba ba ba/, with primary stress on a different syllable each time. With five trained judges to provide the perceptual judgments and a male and a female speaker of general American dialect, the correlation coefficients between the computer estimated and perceptual judgments of stress were 0.913 and 0.903, respectively.

In Experiment II, two passages of connected discourse were chosen. The first passage was chosen because it was reasonably short. The first sentence was relatively simple, and the second sentence showed a somewhat more complex grammatical structure.

As in Experiment I, two speakers, one male and one female, with general American dialects, were asked to read aloud these passages. The recorded tapes were spliced at phrase group boundaries. Rating sessions were conducted with a total number of seven listeners, all of whom spoke with the general American dialect. Perceptual stress judgments were obtained by averaging responses from the seven judges and computer estimated values of stress were obtained.

Table 1 shows the relative contributions of each of the five acoustic parameters to the computer estimates of stress for both experiments. Obviously, the values in Experiment II are lower than those of Experiment I. Local fundamental frequency deviation still remains a strong predictor of stress. However, syllable duration lost some of its significance. The correlation values of the syllable duration variable with perceived stress declined from the range of 0.613 to 0.827 in Experiment I to 0.271 to 0.489 in Experiment II. This drop reflects a much bigger variance in the duration data for connected speech than for nonsense utterances. The correlation of intonational inflection with perceived stress also dropped considerably. This indicates that for normal continuous speech, such inflections were less frequently used.

Table 1 also shows the degree of correlation between the perceptual data and the computer-estimated values of stress. The correlation values for both subjects lie between 0.678 and 0.807, indicating that the computer program provided a less accurate estimation of perceived stress than in Experiment I. The less accurate estimation of stress indicates that there must be additional cues, either linguistic or acoustic, that the listeners used in assigning perceptual stress ratings that were not accounted for by the five acoustic parameters.

V. ISOLATED WORD RECOGNITION AND APPLICATIONS

The recognition accuracy of systems that deal with isolated words or phrases, instead of continuous speech, can be as high as 99%, and there are several commercially available systems that achieve this. This accuracy is sufficient to make such systems very useful in machine-control applications where the hands are busy, and it promises to lead to greatly improved locomotion and environmental control devices for the handicapped.

One such system costs $14,000 with various added options. It uses a 16k, 16-bit microprocessor for processing and an extensive set of features to represent word prototypes and incoming words. It allows a 64-word extendable vocabulary (up to 256 words) for different tasks, with word subset selection. (This improves accuracy and allows the vocabulary to be extended.)

During the set-up phase, each user repeats each chosen word ten times, with a pause of 1/10 sec between each word, and an averaged binary feature set is created for the word. The acoustic signal corresponding to the word is divided into 16 periods, and features such as relative formant frequencies, bandwidths, pitch period, and intensity are thresholded and encoded into 32 bits for each time segment. (Thus, each word is encoded into 512 bits.) During use, the feature set of the incoming word is matched against all of the stored prototype feature sets, and the closest one is selected.

Another system has eight channels at a cost of $60k — 100k (i.e., $7,500 to $12,500 per channel). It is a "speaker-independent" interactive system intended for telephone banking and credit card applications. There is a limited vocabulary of digits and words (usually about 100 words). Each word is encoded into 6144 bits (as compared with 512 bits for the first system), and the system is "speaker-independent" in the sense that it

Table 1
CORRELATION OF ACOUSTIC CUES

	Experiment I (nonsense syllables)		Experiment II (connected speech)	
	Speaker I	Speaker II	Speaker I	Speaker II
Fundamental frequency	0.765	0.820	0.404	0.672
Intensity	0.660	0.804	0.393	0.559
Syllable duration	0.827	0.613	0.271	0.489
Intonational inflection	0.672	0.000	0.278	0.000
Local fundamental frequency deviation	0.793	0.862	0.544	0.666
Computer estimated values	0.913	0.903	0.678	0.807

speaks back to the user, first making the best guess and using "YES/NO" responses to zero-in on the correct choice. However, the system quickly adapts its prototypes to each user and so is useful for telephone banking. A "high" recognition accuracy rate is claimed but is difficult to assess because of the interactive correction and adaptation method used.

A third, less expensive system costing $3,500 uses a single filter and achieves a filter-bank effect by running the signal through the filter at 16 different rates. A reliability rate of 97% is claimed with a vocabulary of 16 words (with "subset" selection allowing 12 different vocabulary sets). Also, three to five repetitions of the user-selected, vocabulary is sufficient.

VI. COMPARISON OF VOICE DATA ENTRY WITH OTHER METHODS

A study made by Welch[16] showed that keyboard input was fastest and most accurate for the entry of numeric data strings by subjects who had keyboard experience, but that it was slow relative to voice and graphical menu for entry of words by inexperienced users with a more complex task.

The principal factors in data entry are

1. Problem complexity and description
2. Data type (strings, scenario, digits or words)
3. Prompting method
4. Feedback effectiveness
5. Degree of hand occupation
6. Operator experience

It has been well demonstrated that for man-to-man communications speech is far superior to other methods, and it now appears that for complex tasks man-computer communication using speech is better than the usual keyboard or menu-pointer methods.

VII. CONTINUOUS SPEECH SYSTEMS

It is much more difficult to recognize phonemes in continuous speech than with isolated words. This is because of coarticulation and segmentation problems. A great deal of progress, however, has been made because of the importance of this area for natural man-machine communication. The well-known ARPA speech project supplied enough funding to several research groups to allow a considerable degree of success.

The Speech Understanding Study Groups at Carnegie-Mellon University, Pittsburgh, produced two of the more successful large-vocabulary, multi-user, continuous, speech understanding systems, viz HEARSAY II[17] and HARPY.[18] These systems will serve to illustrate this area.

A. Evaluation

The ARPA Study Group set goals for the speech project when it began in 1971[19] and listed the factors related to the performance of systems. The following is a slightly modified version of these factors.

Criterion	HARPY performance
Continuous speech vs. isolated words or phrases	(HARPY accepts continuous speech)
Cooperative vs. uncooperative speakers	(Cooperative speakers of general American dialect)
Number of different speakers	(Five speakers)
Quiet vs. noisy environment	(In a computer terminal room)
Quality of microphone and acoustic system	(With an ordinary microphone)
Degree system tuning for each speaker	(Substantial tuning of the system for each speaker, i.e., 25 utterances/speaker)
Degree of user training and adaptation	(Requires no training)
System adaptation to user	(None)
Size of vocabulary	(1011 words with no post selection)
Degree of expandability to larger vocabulary	(Moderately expandable)
Freedom of user to select vocabulary	(None)
Syntactic constraints of chosen languages	(Highly constrained combined semantics/syntax)
Semantic constraints of chosen languages	(Highly contrained combined semantics/syntax)
Model of user vs. no model	(No user model)
Degree of interaction with user	(Modest interaction)
Average recognition accuracy	(9% sentence error — 5% semantic error)
Size of computer memory required	(256 k 36 bits words)
Processor speed required	(0.35 M.I.P.S., PDP-10)
Real time vs. slower response	(80 times real time)
Complexity of software	(Moderate complexity)
Cost per user channel	($5.00 per sentence

Note: The comments in parenthesis show the performance of the HARPY system.[18]

B. Speech Understanding Systems

The HEARSAY II and HARPY systems are quite different. The former uses multiple parallel processors which share a common memory (blackboard), with each processor being responsible for one level of processing, and with the system driven by the state of the common memory. HARPY has a simpler approach using a state transition network with data-dependent transition probabilities, heuristically modified dynamic programming, and segmentation. HARPYs success is mainly due to the task-oriented grammar and the global scoring scheme associated with dynamic programming. In contrast, the HEARSAY II system operates somewhat like a model of human speech understanding. The blackboard (common memory) is organized as via multi-level data structure with various levels communicating with each other, the individual processors (knowledge sources). The levels are (1) parametric, (2) segmental, (3) phonetic, (4) surface-phonemic, (5) syllabic, (6) lexical, (7) phrasal, and (9) conceptual.

The function of the knowledge sources can briefly be described as follows:

Segmentation — Silence is first separated from nonsilence, then the nonsilent region is broken down into sonorant and nonsonorant regions. Eventually, five different types of segment are produced; namely, silence, sonorant peak, sonorant nonpeak, fricative, and flap. The duration is associated with each segment with its absolute amplitude, and amplitude relative to its neighboring segments (i.e., local peak, local value, or plateau). Finally, segments are labeled with allophonic designators (98 different kinds). This KS transfers information from the first to the second level.

Word-spotting — The initial generation of a word is done in a bottom-up fashion using data from the segmental level. Three KSs are used here, namely, "POM, MOW, and WORD-CTL". The POM KS generates a hypothesis for likely syllable class, the MOW KS looks up each hypothesized syllable class and generates word candidates from the set of words containing that syllable class, and is controlled by WORD-CTL based on the number of hypotheses. Each word generated has its begin and end time and confidence score attached.

Word island generation — The Word-Seq KS generates, bottom-up, from the word hypotheses, a small set (3 to 10) of word sequences. It takes the highest rated single words and generates multi-word sequences, by expanding them with other hypothesized words that are both time and language adjacent. The Word-Seq-Ctl KS controls the number of hypotheses created in Word-Seq.

Word sequence parsing — The parse KS creates a word sequence of arbitrary length, using the full constraints of the language.

Word predictions from phrases — The predict KS generates predictions of all words which can immediately precede or follow a given phrase in the language.

Word verification — The verify KS verifies the existence of or rejects, each such predicted word in the context of its predicting phrase. It also checks for time adjacency.

Word phrase concatenation — This KS generates longer phrases from the verified predicted phrases.

Complete sentences and halting criteria — The stop KS prunes the remaining candidates; if the completed sentence has a higher rating than that expected of remaining sentences, and the process halts.

The Hearsay II system came close to meeting the goals of the ARPA Project with 10% semantic error in test utterances and 17% error in word recognition. The use of a best-first strategy allowed the system to follow one syntactic path to completion very quickly, but because of the need for back-tracking, and the search time-limit cut-off, some utterances were missed.

REFERENCES

1. Reddy, D. R., Erman, L. D., Fennell, L. D., and Neely, R. B., The 'Hearsay' Speech Understanding System: An Example of the Recognition Process, Proc. 3rd Int. Jt. Conf. on Artificial Intelligence, Stanford, Calif., August 20 to 23, 1973.
2. Holden, A. D. C. and Strasbourger, E., A computer programming system using continuous speech input, *IEEE Trans. ASSP,* ASSP-24(6), 579, 1976.
3. Atal, B. S. and Hanauer, S. L., Speech analysis and synthesis by linear prediction of the speech wave, *J. Acoust. Soc. Am.,* 50, 637, 1971.
4. Itakura, F. I. and Saito, S., Analysis-Synthesis Telephony Based Upon the Maximum Likelihood Method, Proc. 6th Int. Cong. Acoustics, Tokyo, 1968, C17.
5. Itakura, F. I. and Saito, S., A statistical method for estimation of speech spectral density and formant frequencies, *Electron. Commun. Jpn.,* 53A(1), 36, 1970.
6. Chandra, S. and Lin, W. C., Experimental comparisons between stationary and non-stationary formulation of linear prediction applied to speech, *IEEE Trans. ASSP,* ASSP-22, 403, 1974.

7. Rabiner, L. R. and Schafer, R. W., *Digital Processing of Speech Signals,* Prentice-Hall, Englewood Cliffs, N.J., 1978.

8. Markel, J. D. and Gray, A. H., Jr., On autocorrelation equations as applied to speech analysis, *IEEE Trans. Audio Electroacoust.,* AU-21(2), 69, 1973.

9. Markel, J. D., The SIFT algorithm for fundamental frequency estimation, *IEEE Trans. Audio Electroacoust.,* AU-20(5), 367, 1972.

10. Cheung, J. Y., Holden, A. D. C., and Minifie, F. D., Computer recognition of linguistic stress patterns in speech, *IEEE Trans., ASSP,* ASSP-25(3), 252, 1977.

11. Cheung, J. Y. and Holden, A. D. C., Computer modeling and estimation of linguistic stress patterns, *IEEE Int. Conf. on Acoust. Speech Signal Process.,* IEEE, Piscataway, N.J., 1976.

12. Lea, W. A., Medress, M. I., and Skinner, T. E., Prosodic aids to speech recognition, Sperry Univac Computer Systems, Rep. PX7940, Sperry Univac, St. Paul, Minn., 1972.

13. Lea, W. A., Medress, M. I., and Skinner, T. E., Prosodic aids to speech recognition: II syntactic segmentation and stressed syllable location, Sperry Univac Computer Systems, Rep. PX10232, Sperry Univac, St. Paul, Minn., 1973.

14. Lea, W. A., Medress, M. I., and Skinner, T. E., A prosodically guided speech understanding strategy, *Proc. IEEE Int. Symp. on Speech Recognition,* IEEE, Piscataway, N.J., 1974.

15. Chomsky, N. and Halle, M., *The Sound Pattern of English,* Harper & Row, New York, 1968.

16. Welch, J. R., Automatic Data Entry Analysis, Tech. Rep. No. RADL-TR-77-306, Threshold Technology, Inc., Delran, NJ, August 1977.

17. Lesser, V., Fennel, R., Erman, L., and Reddy, D. R., Organization of the Hearsay II Speech Understanding System, *IEEE Trans. Acoust. Speech Signal Process.,* ASSP-23, 11, 1975.

18. Lowerre, B. T., The HARPY Speech Recognition System, Tech. Rep. Computer Science Dept., Carnegie-Mellon University, Pittsburgh, Pa, April 1976.

19. Newell, A., Barrett, J., Forgie, J. W., Green, C., Klatt, D., Licklider, J. C. R., Munson, J., Reddy, R., and Woods, W., Speech Understanding Systems: Final Report of a Study Group Tech. Rep., Computer Science Dept., Carnegie-Mellon University, Pittsburgh, Pa, May 1971.

Chapter 7

APPLICATIONS TO GEOPHYSICS: PART I

C. H. Chen

TABLE OF CONTENTS

I. INTRODUCTION

In recent years, computers have played an increasingly important role in geophysics, especially in seismic studies such as petroleum exploration, nuclear detection, earthquake research, and marine seismic studies. Computers are needed to process large volumes of seismic data from which useful information must be extracted accurately. In some geophysical applications, only a limited amount of data is available but a high resolution spectral analysis is needed, which also requires a great deal of computation. The geophysics area is indeed one of the largest users of computer resources. Proper choice of seismic signal processing techniques is thus extremely important. In this chapter, we provide an overview of seismic signal processing, which is followed by a detailed study of computer-aided discrimination of nuclear explosion and natural earthquake events and a brief discussion on earthquake prediction and mineral extraction. In the following chapter, the problem of intrusion detection via adaptive filtering and a high resolution maximum entropy method for geophysics application will be examined.

II. SEISMIC SIGNAL MODELING

A good understanding of the signal generation process along with the mathematical signal modeling is usually needed for effective signal processing. Figures 1, 2, and 3 depict simplified transmission processes of seismic waves. Figure 1 shows an inhomogeneous earth excited by a deep source. The earth is bounded by two homogeneous infinite half-spaces, the air and the basement rock. Here the earth is a distributed parameter system governed by partial differential equations. For digital processing, the originally continuous velocity profile can be quantized and, as a result, the earth can now be modeled as a lumped parameter system. If the time of signal propagation through a layer is short compared with the duration of the signal, then the lumped parameter assumption is valid. By choosing the depth of each layer to be very small, i.e., considering many layers, we can satisfy the lumped parameter conditions.

To simplify the analysis, we can assume that the system of Figure 1 is linear and time invariant. Let a represent the constant parameters associated with the N layers. We can expect the input x_n and the output y_n to satisfy the following linear difference equation:

$$y(n) + a_1 \, y(n-1) + \cdots + a_N y(n-N) = x(n) \qquad (1)$$

Taking Z-transform on both sides of Equation 1, we obtain the ratio

$$\frac{Y(z)}{X(z)} = \frac{1}{1 + a_1 \, z^{-1} + a_2 \, z^{-2} + a_N z^{-N}} = B(z) \qquad (2)$$

which represents the transfer function of an all-pole filter. Our physical claim that the lumped parameter model represents a stable system is equivalent to the mathematical condition that the transfer function contains no poles outside the unit circle. Equation 1 also represents an autoregressive model for the digitized seismic signal.

Figure 2 describes the transmission of teleseismic waves, which gives a better physical insight of the seismic signal generation. For mathematical analysis, a simplified model such as Figure 3 is more appropriate. It shows internal primary reflections caused by a down-going unit impulse δ(n) applied at the surface. For clarity, ray paths are drawn

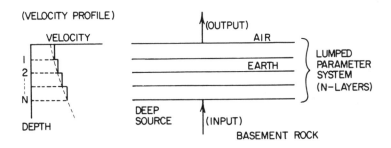

FIGURE 1.   An inhomogeneous earth excited by a deep source.

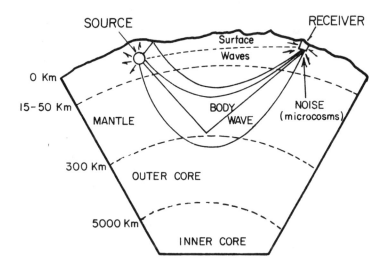

FIGURE 2.   Transmission of teleseismic waves.

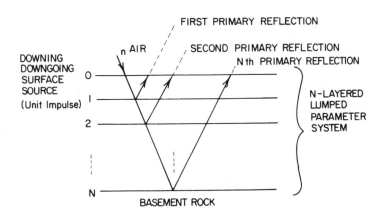

FIGURE 3.   A simplified model of Figure 1.

at oblique incidence, but wave motion analysis is for normal incidence. Let b(n) be
the response of a single layer with respect to the unit impulse input δ(n). By linearity
property, a delayed impulse δ(n-m) gives rise to b(n-m) and cδ(n) gives rise to cb(n),

where c is a constant. By superposition principle, the total impuse response can be written as

$$h_n = \epsilon_1 b(n-1) + \epsilon_2 b(n-2) + \cdots + \epsilon_n b(n-N) = \epsilon_n * b(n) \qquad (3)$$

where "*" denotes convolution and ϵ_n are the hypothetical sources of strength given by the reflection coefficients γ_n at various layers. Taking the z-transform of Equation 3, we have

$$H(z) = E(z) B(z) = \frac{\epsilon_1 z^{-1} + \epsilon_2 z^{-2} \cdots + \epsilon_n z^{-n}}{1 + a_1 z^{-1} + a_2 z^{-2} + \cdots a_n z^{-n}} \qquad (4)$$

which is the transfer function of a normal-incidence reflection seismogram. It is also called the ARMA (autoregressive moving average) model used in the reflection seismology.

Since the layered system is assumed to be both linear and time invariant, the reflection seismogram y(n) due to an arbitrary source pulse s(n) is

$$y(n) = h(n) * s(n) = \epsilon_n * b(n) * s(n)$$

$$= \epsilon_n * (b(n) * s(n)) = \epsilon_n * w(n) \qquad (5)$$

where w(n) = b(n) * s(n) is defined as a composite wavelet consisting of the reverberation wavelet b(n) and source pulse s(n). Thus, Equation 5 describes the normal incidence reflection seismogram, where b(n) represents the autoregressive component and s(n)*ϵ_n the moving average component. The above discussion illustrates the seismic signal generation as a convolution process between the source and the impulse response of the earth. A basic deconvolution problem in signal processing is to filter y(n) such that we can best recover the reflection coefficient sequence ϵ_n.

III. SPECTRAL MATCHING AND THE ARMA MODEL

Figure 4 shows a typical set of teleseismic waves studied. The data are digitized at ten samples per second. The Fourier transform of 1024 data points is also shown along with typical explosion records, (A) and (B), and typical earthquake records, (C) and (D). The magnitude spectrum may be smoothed by the modified periodogram[1] and other techniques.

The ARMA model, as derived in the previous section, is the most general linear seismic signal model that provides simple parametric representation of the signals. Theoretically speaking, the spectrum of any physical signal can be matched, i.e., fitted, perfectly by an autoregressive model with an arbitrarily high order. For a typical set of teleseismic waveforms, a good spectral matching based on the autoregressive, i.e., all-pole, model has been reported.[2,3] If the ARMA, i.e., pole-zero, model is used, a better spectral matching is available. The degree of spectral matching can be measured by the mean squared error between the actual and the modeled signals. Figures 5 to 7 illustrate spectral matching by both all-pole and pole-zero models. The pole-zero models have lower mean squared errors that the corresponding all-pole models. The 15th order models show a better spectral fit than the corresponding 9th order models. For more complicated spectra as shown in Figure 7, even the 15th order model would have difficulty in providing a good spectral match. However, the pole-zero model is

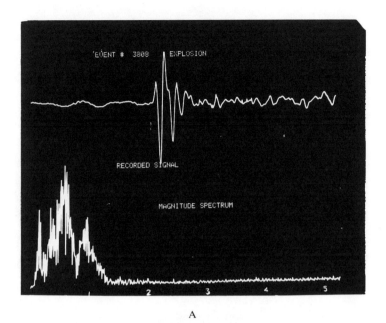

A

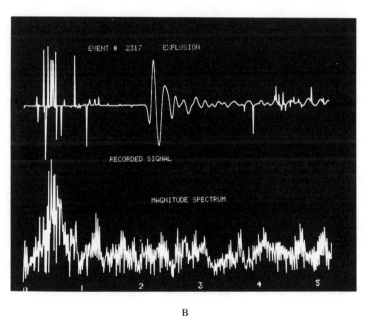

B

FIGURE 4.   A typical set of seismic waves; (A) (B) explosions, (C) (D)
earthquake records.

superior to the all-pole model. Generally speaking the problems associated with the
ARMA model are as follows: (1) The order of the model must be finite in practice.
The linear model is limited in its capability not only in spectral matching but also in
representing only a first order approximation to the original signal. (2) Computationally, the order of the model and the coefficients in both the numerator and denominator polynomials must be determined. This is not a simple task and has been an area
of active research.

For the all-pole, i.e., autoregressive, model, the determination of parameters has

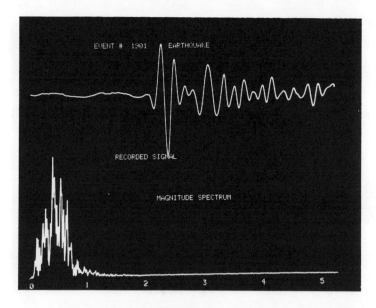

FIGURE 4C

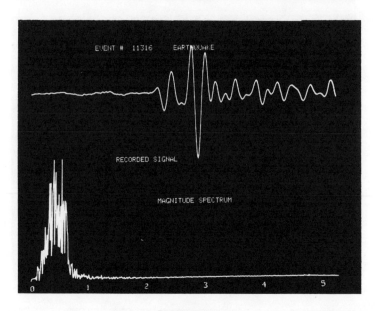

FIGURE 4D

been discussed in Chapter 3. The computer results presented in Figures 5 to 7 are based on the Levinson recursion algorithm. The computer program listing of the algorithm is included in the Appendix of this chapter. After the all-pole model is determined then a numerator polynomial is added. Coefficients of the numerator are adjusted, without changing the denominator, to minimize the mean squared error between the data point and the linearly predicted value. Even though the results shown in Figures 5 to 7 as pole-zero models are suboptimal, the procedure is computationally efficient.

The application of ARMA models to seismic signal processing is not limited to spectral estimation and data compression. There are good physical interpretations of parameters and related quantities such as the reflection coefficients. The parameters are potentially useful features for classification of teleseismic events. Good spectral esti-

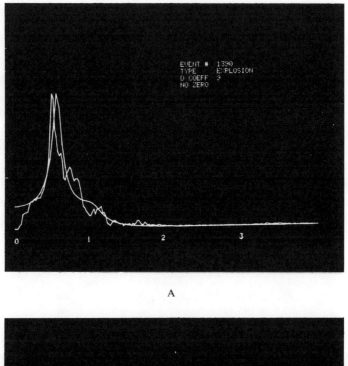

A

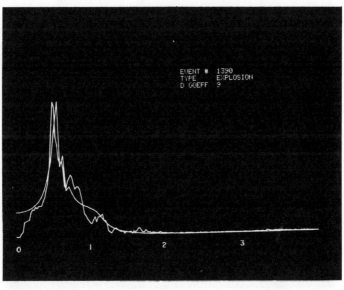

B

FIGURE 5. Spectral matching for an explosion spectrum by using; (A) 9th order all-pole model; (B) 9th order pole-zero model; (C) 15th order all-pole model; and (D) 15th order pole-zero model.

mation leads to accurate computation of spectral ratio, which is another useful feature in seismic discrimination.

IV. PREDICTIVE VS. HOMOMORPHIC DECONVOLUTION

In determining the source pulse by using the deconvolution method, both the predictive and homomorphic deconvolution methods have been extensively studied. They represent two important but quite different approaches to the problem.[4] In Equation

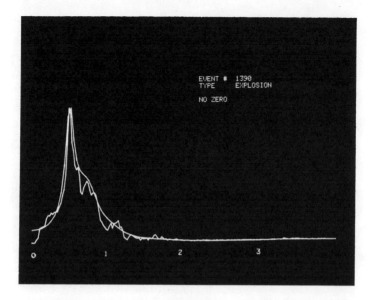

FIGURE 5C

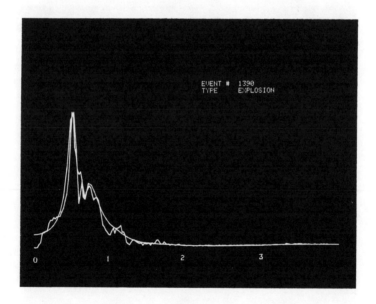

FIGURE 5D

5, the theoretical reverberation wavelet b(n) is minimum phase, while the source pulse s(n) is not. For a given signal autocorrelation, only the minimum phase pulse denoted as s_0(n) corresponding to s(n) can be determined. We can write

$$s(n) = s_0(n) * \rho(n) \qquad (6)$$

where ϱ(n) is the all-pass filter. The problem in predictive deconvolution is thus to determine an all-pass filter ϱ(n) to obtain a source pulse with correct phase characteristic. If we know ϱ(n) and thus the inverse filter ϱ^{-1}(n), then

$$\rho^{-1}(n) * y(n) = \epsilon_n * (b(n) * (b(n) * s_0(n))) \qquad (7)$$

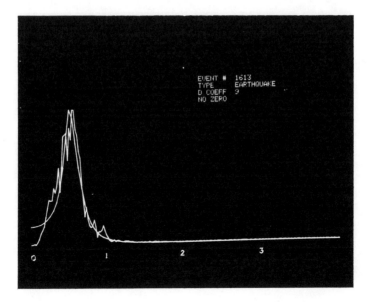

A

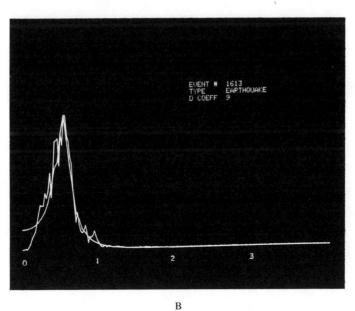

B

FIGURE 6. Spectral matching for an earthquake spectrum by using; (A) 9th order all-pole model; (B) 9th order pole-zero model; (C) 15th order all-pole model; and (D) 15th order pole-zero model.

and thus the method of predictive deconvolution will perform optimally. In practice however, $\varrho(n)$ is not known and we have as yet no truly general technique to estimate the appropriate $\varrho(n)$ from $y(n)$.

The complex cepstrum of $y(n)$, denoted as $\hat{y}(k)$, is defined as the inverse Fourier transform of the logarithm of the Fourier transform of $y(n)$. Thus

$$\hat{y}(k) = \hat{\epsilon}_k + \hat{b}(k) + \hat{s}(k) \tag{8}$$

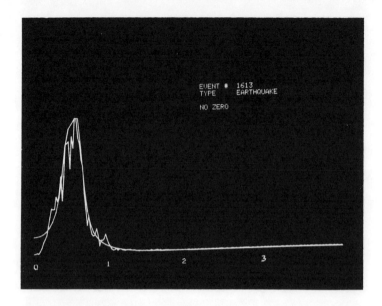

FIGURE 6C

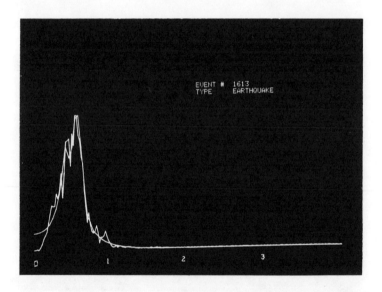

FIGURE 6D

where k is the guefrency. The frequency domain is defined for both positive and negative k. The minimum phase sequences have only positive frequency components, while maximum phase sequences have only negative frequency components. Thus, s(n) is generally not minimum phase but rather a mixed phase, and ε_n should also be considered as mixed phase because the reflection coefficient series ε_n is uncorrelated and random. Thus, the complex cepstrum $\hat{y}(k)$ will have both positive (minimum phase) and negative (maximum phase) frequency components. The low-frequency portion of the cepstrum often can be identified with the cepstrum of the source pulse $\hat{s}(k)$. In principle, this suggests that the source pulse $\hat{s}(n)$ can be recovered from the "low-pass", also called "short-pass", portion of $\hat{y}(k)$. In practice there is no perfect cutoff frequency, and some trial-and-error is needed to obtain a reasonable estimate of the

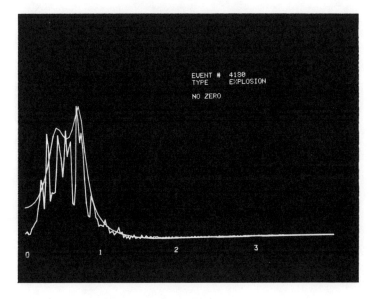

A

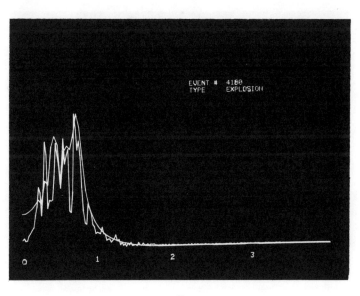

B

FIGURE 7. Examples of more complicated spectra. Spectra matching for
an explosion spectrum by using; (A) 15th order all-pole model and (B) 15th
order pole-zero model. Also, spectral matching for an earthquake spectrum
by using (C) 15th order all-pole model and (D) 15th order pole-zero model.

source pulse. By the same token we can use trial-and-error to estimate ϱ_n in the predictive deconvolution method.

Although the assumption about delay (or phase) properties of the source pulse is essentially required in both methods, such assumption is less critical in homomorphic deconvolution than in predictive deconvolution. In homomorphic deconvolution an exponential weighting of the data sequence is usually necessary to remove the computational instability due to the nonminimum phase source pulse. Figure 8 shows some

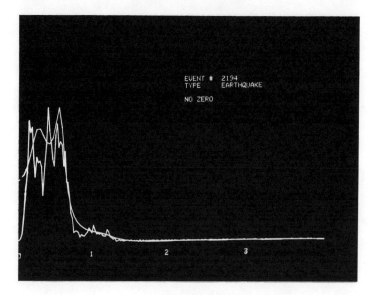

FIGURE 7C

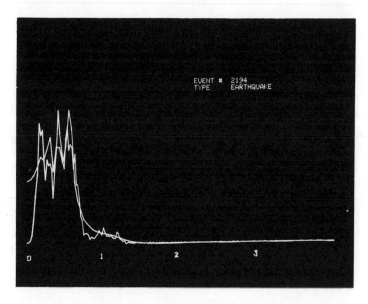

FIGURE 7D

results of cepstral analysis on the teleseismic records. There has been more success with predictive deconvolution than with the homomorphic deconvolution in petroleum exploration. Further work on the homomorphic deconvolution,[5] however, can make the method more feasible.

V. GEOPHYSICAL FEATURES FOR SEISMIC DISCRIMINATION

Modeling and processing described in the previous sections are essential steps to extract useful information from the seismic data. Interpretation of teleseismic events by man or machine or both is another important step in seismic signal study. Human interpretation based on the waveforms and the spectra is usually unreliable unless ad-

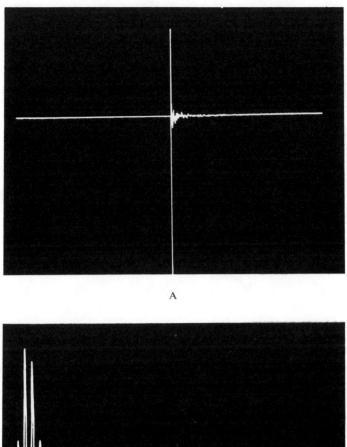

A

B

FIGURE 8.   Cepstral analysis of a typical earthquake record. The data are
exponentially weighted with $\alpha = 0.998$.[1] (A) The complex cepstrum, (B) low-
pass output with window size of 30, and (C) high-pass output with window
size of 30.

ditional information, such as the location of the source, is available. Automatic interpretation requires properly defined features. Clearly the seismic waves are governed
by geophysical phenomena. It is desirable and even necessary to extract features with
geophysical significance, based on which the seismic events can be classified. With the
use of computer-aided analysis techniques, the seismologists during the past 15 years
have discovered a number of effective discriminants such as the following:[6,7]

1.   The observed differences in the P-wave amplitude spectra, especially at periods

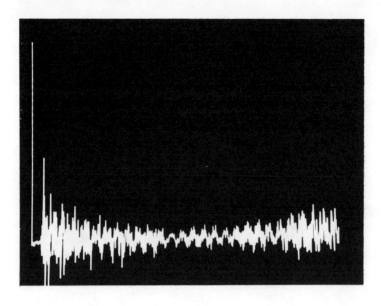

FIGURE 8C

longer than 3 sec. The explosion spectra peak sharply in the interval between 1 and 2 sec and decay rapidly for longer periods. The earthquake spectra, however, continue to increase to a long-period interval. Note that frequency is the inverse of period.

2. The surface waves for long-period records. For a given bodywave magnitude, surface wave magnitudes are generally smaller for underground explosions than for earthquakes.

3. The amplitude spectra of Rayleigh waves for detecting underground explosions from small earthquakes.

4. The amount of P and S waves. Earthquakes produce approximately equal amounts of P and S waves, while explosions produce more P waves.

5. The first onsets. Earthquakes give anaseismic and kataseismic first onsets, and explosions give anaseismic first onsets everywhere.

6. The focal depth. Earthquakes have relatively deep foci, while explosions have only shallow foci.

7. The duration. Earthquakes generally have longer durations than explosions.

8. The ratio of surface wave magnitude to bodywave magnitude.

There are other useful geophysical discriminants. Unfortunately, the direct application of these discriminants is not feasible. Although explosions provide more P waves than S waves, not all earthquakes follow the average pattern. Further, it is very difficult to identify particular phases for low magnitude earthquakes. The difficulty in determining the direction of onset, due to noise, obscures the anaseismic-kataseismic discrimination feature. Focal depth holds a great deal of promise. However, there is still too much error associated with measurement to warrant its exclusive use. Again, the signal-to-noise ratio is too small, for small-magnitude events, to discriminate using the duration of the wave trains. For automatic classification, the discriminants or features must be computable with high accuracy. Let $y(t)$ be the seismic signal trace with corresponding Fourier transform $Y(f)$. Three useful discriminants are complexity (C), spectral ratio (SR), and third moment of frequency (TMF). They have been empirically determined as reliable features with the following definitions:[7,8]

$$C = \int_0^5 y^2(t)dt / \int_5^{35} y^2(t)dt$$

$$SR = \int_{L1}^{L2} |Y(f)| df / \int_{H1}^{H2} |Y(f)| df$$

where the limits of the low and high frequency bands in Hertz have been selected as

	L1 L2	H1 H2
U.S. events	0.55—0.94	1.48—1.87
U.S.S.R. events	0.66—0.94	2.19—2.89

$$TMF = \left[\int_{0.32}^{5.0} |Y(f) - X(f)| f^3 df / \int_{0.32}^{5.0} |Y(f) - X(f)| df \right]^{1/3}$$

where X(f) is the spectrum of the instrument noise alone. For digitized data, the integration is replaced by summation. Among the three features, it is our experience that the complexity is most effective, the TMF ranks second, while the SR is less effective than the other two. The spectral ratio is affected by the noise spectrum especially when the signal is weak. For noise alone the spectral ratios were found to be 3.6 for the U.S.S.R. events and 3.3 for U.S. events.

VI. COMPARISON OF FEATURE EXTRACTION CRITERIA

As a number of feature extraction criteria are available, comparative evaluation of such criteria is necessary. The comparison will be based on the performance (percentage correct recognition) and the computational requirements. Two expanded seismic data bases were used for our recognition study. The first set has 157 earthquakes and 157 explosion records. The second set has a multichannel data base with 214 earthquake and 113 explosion events. The seismic events all recorded at the LASA (large-aperture seismic array) facility in Montana have a wide variety of source origins, including central Asia, Turkey, western China, etc. These are the most representative seismic records available for classification study.

The three simple discriminants described in the previous section are easy to compute. The use of a single discriminant provides a low percentage correct recognition (70% or less). The use of all three discriminants does not improve the performance. The best combination of two discriminants is the complexity and the third moment of frequency, leading to only 80% correct recognition with our data base. Thus, these discriminants are inadequate for practical classification purposes.

The complex cepstrum provides useful features with good recognition result (94.4% correct) based on a set of 36 seismic records.[9] Only 72% correct recognition is available by using the third moment of frequency and spectral ratio for the same records. Features may be derived from orthogonal transforms, but a large number of transform coefficients are needed for a meaningful performance. The Karhunen-Loeve transform method can be used for feature compression. Statistical techniques are very useful for seismic discrimination. By considering the same wave as a stationary Gaussian process, the divergence and linear decision function have been used for classifying a small number of seismic records.[10] In practice the seismic wave is not strictly stationary; the spectral averaging over several time intervals can provide more effective mathematical features, such as the dynamic spectral ratio[11] and the short-time spectral ratio.[9,12]

The autoregressive moving average (ARMA) model coefficients can be used as features for a single channel waveforms described above.[9,13] It is also proposed that the autoregressive coefficient matrices appearing in a multivariate autoregressive model fitting may be used for feature extraction purposes in problems concerning the recognition of multichannel waveforms.[14] The autoregressive parameters may be further compressed by applying the Karhunen-Loeve transform method. The results obtained by using a multivariate Gaussian classification suggest that the combined autoregressive/Karhunen-Loeve transform method has a considerably larger discrimination potential than the more conventional seismic discriminants described earlier in this chapter.

Assume that $[Y_1, \ldots, Y_q]$ is a sample from a q-dimensional zero mean, real-valued weakly stationary process $\{Y(n)\}$. Then the linear difference equation for the multichannel time series can be written as[15]

$$Y(n) + A_1 Y(n-1) + \cdots + A_N Y(n-N) = W(n) \qquad (9)$$

where $A_1, ---, A_N$ are q by q matrices and $W(n)$ is a multichannel white noise series. The coefficients can be estimated by solving the matrix Yule-Walker equation.[15] The feature vector obtained from the coefficient matrices $A_1, ---, A_N$ and the covariance matrix of $W(n)$ usually has a very high dimension, and thus, feature space compression is needed.[14]

VII. SEISMIC DISCRIMINATION USING DECISION THEORY

After proper features are selected, classification can be performed. Because of the difficulty in detecting surface waves from weak explosions, most discriminants constructed by seismologists are based on body-wave or short-period information. Only linear decision functions are used. Better utilization of the statistical decision theory can lead to better classification performance. Fisher's linear discriminant, maximum likelihood decision rule and the nearest neighbor decision rule, as described in Chapter 4, are most typically used. For the maximum likelihood decision rule, a probability density must be assumed or estimated. The multivariate Gaussian density is often assumed. Let $Z = [Z(1), ---, Z(d)]$ be the compressed feature vector corresponding to coefficient matrices in connection with Equation 9. The Gaussian probability density for the ith class (category) is

$$p(z/\omega_i) = (2\pi)^{-d/2} |K_i|^{-\frac{1}{2}} \exp\left\{ -\tfrac{1}{2}(z-\mu_i)' K_i^{-1} (z-\mu_i) \right\}; \quad i=1,---, m \qquad (10)$$

where μ_i and K_i are the mean vector and covariance matrix of the ith class. The maximum likelihood decision rule assigns z to the class with the largest $\varrho(z/\omega_i)$. The pattern class in this case refers to the category of seismic events.

If the available number of training samples is small and there is not sufficient confidence on the Gaussian assumption, the nearest neighbor decision rule should be used. Computational algorithms are available to reduce the amount of computation and storage with this nonparametric decision rule. The recognition results of the maximum likelihood decision rule with properly chosen probability density and the nearest neighbor decision rule are very much the same in practice for a limited number, say 100, training samples. Without feature space compression, slightly over 90% correct recognition has been available by using the nearest neighbor decision rule for our data bases.[9,16] The features considered are the prediction coefficients of an all-pole model and the autocovariance features. Generally speaking, there are more incorrect decisions for explosions than for earthquakes.

Because of the wide variety of source origins for both earthquakes and explosions, a very large data base is needed to establish reliable statistical information for feature computation and for good recognition accuracy. Effort in this direction is greatly needed for automatic discrimination of seismic events occurring throughout the world.

VIII. EARTHQUAKE PREDICTION

Human prediction of earthquakes with simple equipment was started in China several hundred years ago. Presently, very sophisticated equipment is used throughout the world to continuously monitor the seismic events for human interpretation and prediction of earthquakes. Computers have made it possible to establish networks of earthquake prediction facilities. Computer-aided earthquake prediction efforts contribute to significant advances in the prediction capability currently available. However, fully automatic earthquake prediction still requires considerable efforts. Continued research is greatly needed in the automatic processing of local and global earthquake data.[17,18]

Microearthquake data are currently being used in earthquake prediction, locating and monitoring geothermal areas, and three dimensional seismology. The use of microearthquakes in earthquake prediction is based on the idea that large earthquakes share the same tectonic causes as the numerous small ones occurring in the same general area. The full exploitation of this data for the purpose of earthquake prediction has been hampered by a lack of a strict, uniform procedure for analyzing microearthquake data. For example, the bias and error in elementary measurements, such as picking the time of first arrivals, may vary from time to time because of a change in personnel hired for the work. The subjective bias in reading the seismograms, especially in picking the arrivals, has an important effect on first motion studies. Often the wrong sense of first motion is picked if the signal to noise ratio is lower than a critical value.

The California seismic networks record an enormous amount of data annually. For example, in 1 year the Southern California network recorded over 260,000 seismograms from over 7000 events, and at least as many seismograms can be expected from the Central California network each year. Even if seismograms can be processed manually at this rate, earthquake swarms and aftershock sequences can increase the seismicity level of an area by an order of magnitude. Rapid, objective analysis of such a volume of data could be crucial for earthquake prediction. Recently, the rate at which seismic data can be analyzed has been vastly increased by the use of the California Institute of Technology Earthquake Detection and Recording (CEDAR) system. CEDAR is a real-time computer system that detects earthquakes on the Southern California array and records the seismograms for later processing. The final output of CEDAR is a data base that includes earthquake locations, P and S wave arrival times, and the seismograms. Thus for the first time, CEDAR provides all of the raw materials that a seismologist needs in a format that is easily manipulated by computer. An excellent discussion has been written by Anderson[17] on how seismologists measure arrival times and locate earthquakes in order to develop an objective and reliable computer program to aid in seismic analysis of the CEDAR data.

There is always ambiguity associated with measuring first arrival time from seismograms whether it is done by a seismologist or by a machine, since the seismic signals are of unknown shape and are contaminated with noise. This ambiguity increases with distance from the epicenter because of the structure and attenuation of the earth. For example, beyond 100 km distance the first arrival becomes very weak and can easily be confused with the much stronger Moho reflection. In order to reduce the effect of this ambiguity, information from many sensors must be used to constrain arrival times on individual sensors. Since one of the most important constraints is the location of the event, the processes of picking arrivals and locating the event should be combined in an iterative fashion.

The first-arrival-time algorithm[17] that makes use of the information feedback provided by the locating stage, can estimate the arrival time accurately. The arrivals are used to determine an initial location for the event. Arrivals that were missed or were inconsistent with this location are repicked using more appropriate parameter settings. The earthquake is then relocated. This process may be repeated until no arrivals need to be repicked.

Briefly, the algorithm for estimating the arrival accuracy is as follows: the first half cycle of the first arrival is represented by four points; the first zero crossing, the first trigger threshold crossing, the first maximum in the half cycle, and the next zero crossing. The standard deviation of the arrival time is given by

$$\sigma_t = \frac{\sigma_n}{M} \qquad\qquad (11)$$

where σ_n is the standard deviation of the noise and M is the slope of the line between the threshold crossing and the maximum. Heuristically speaking, the equation says that for identical arrivals, the accuracy is determined by the noise level, and for the same noise level, sharper arrivals should be more accurate than weaker ones. This is in agreement with the normal practice of weighting impulsive arrivals higher than emergent ones in an earthquake location.

IX. MINERAL EXTRACTION

The use of seismics in mining exploration has been restricted so far to only some trial measurements. If seismic methods come into more prominence in ore exploration, as they most certainly will do in the future, we can be sure that the signals of interest will be hidden far more "deeply" than is the case for the signals in oil prospecting. Some very sophisticated data processing will be required to extract them. The highly developed reflection seismology that reveals subsurface structure information will be useful to mineral extraction. Further advances in geophysical signal processing are definitely needed for an effective computer-aided mineral extraction.

APPENDIX

PROGRAM MARKEL FOR COMPUTATION OF AR MODEL COEFFICIENTS

NAME - MARKEL

Subroutine MARKEL (Temp1,C,NC,ERRN)

```
C
C       Temp1 — Autocorrelation array
C       C — Linear predictor coefficients
C       NC — Number of coefficients
```

```
C          ERRN — Normalized error
C

           Dimension Temp1(1),C(1)
           Dimension Temp2(51),Temp3(51)
           IF(NC. LE. 51)GO to 1
           NC = 51
C
C          Generating the coefficients
C
1          Temp2(1) = 1
           RO = Temp1(1)
           Alpha = Temp1(1)
           Beta = Temp1(2)
           C(2) = -Beta/Alpha
           Gamma = C(2)*Beta
           NCC = NC−1
           DO 11 N = 2,
           NM1 = N−1
           A = −Beta/Alpha
           IF(N−2)7,7,4
4          DO 5 J = 2, NMI
           NN = N−J + 1
5          Temp3(J) = Temp2(J) + A*Temp2(NN)
           DO 6 J = 2, NM1
6          Temp2(J) = Temp3(J)
7          Temp2(N) = Temp2(1)*A
           Alpha = Alpha + A*Beta
           Beta = 0
           DO 8 J = 1, N
           NN = N−J + 2
8          Beta = Beta + Temp2(J)*Temp1(NN)
           Q = −(Temp1(N + 1) + Gamma/Alpha
           DO 9 J = 1, NM1
           NN = N−J + 1
9          C(J + 1) = C(J + 1) + Q*Temp2(NN)
           C(N + 1) = Q*Temp2(1)
           Gamma = 0
           DO 10 J = 1, N
           NN = N− J + 2
10         Gamma = Gamma + C(J + 1)*Temp1(NN)
11         Continue
           C(1) = 1
C
C          Calculates the normalized error
C
           SS = 0
           DO 12 J = 1, NCC
12         SS = SS + C(J + 1)* Templ(J + 1)
           ERRN = 1 + SS/RO
           Return
```

BIBLIQGRAPHIC NOTES

There is a large amount of literature on computer-aided teleseismic discrimination and oil exploration. In addition to those references cited above, the Proceedings of the 1977 Institute of Electrical and Electronics Engineers International Symposium on Computer-Aided Seismic Analysis and Discrimination has an extensive list of references on teleseismic discrimination. Besides the journals such as *Geophysics* and *Bulletin of Seismological Society of America* and others, many recent publications[9-24] are very important sources of reference for the rapid development in geophysical signal processing and geophysical time series analysis.

REFERENCES

1. Oppenheim, A. V. and Schafter, R. W., *Digital Signal Processing*, Prentice-Hall, Englwood Cliffs, N.J., 1975, chap. 11.
2. Chen, C. H., On Digital Signal Modelling and Classification with the Teleseismic Data, IEEE Int. Conf. on Acoustics, Speech and Signal Processing Record, Tulsa, Okla., April 1978.
3. Chen, C. H., Recognition of Teleseismic Waveforms, Proc. Int. Comput. Symp. Taiwan, December 18 to 20, 1978.
4. Treitel, S. and Robinson, E. A., Deconvolution-homomorphic or predictive, *IEEE Trans. Geosci. Electron.*, GE-15(1), 11, 1977.
5. Tribolet, J. M., *Seismic Applications of Homomorphic Signal Processing*, Prentice-Hall, Englwood Cliffs, N.J., 1979.
6. Dahlman, O. and Israelson, H., *Monitoring Underground Nuclear Explosions*, Elsevier, Amsterdam, 1977.
7. Båth, M., Spectral Analysis in Geophysics, Elsevier, Amsterdam, 1977.
8. Chen, C. H., Feature extraction and computational complexity in seismological pattern recognition, *Proc. 2nd Int. Jt. Conf. on Pattern Recognition*, IEEE, Piscataway, N.J., 1974.
9. Chen, C. H., Seismic pattern recognition, *Geoexploration J.*, 16(½), 133, 1978.
10. Schumway, R. H. and Unger, A. N., Linear discriminant functions for stationary time series, *J. Am. Stat. Assoc.*, 67, 948, 1974.
11. Ives, R. B., Dynamic spectral ratios as features in seismological pattern recognition, *Conf. on Comput. Graphics, Pattern Recognition and Data Structure*, IEEE, Piscataway, N.J., 1975.
12. Chen, C. H., Application of pattern recognition to seismic wave interpretation, in *Applications of Pattern Recognition*, Fu, K. S., Ed., CRC Press, Boca Raton, Fla., in press.
13. Tjøstheim, D., Recognition of waveforms using autoregressive feature extraction, *IEEE Trans. Comput.*, C-26, 268, 1977.
14. Tjøstheim, D. and Sandvin, O., Multivariate autoregressive feature extraction and the recognition of multichannel waveforms, *IEEE Trans. Pattern Anal. Mach. Intelligence*, PAMI-1(1), 80, 1979.
15. Robinson, E. A., *Multichannel Time Series Analysis with Digital Computer Programs*, Holden-Day, San Francisco, 1967.
16. Chen, C. H. and Lin, I. C., Pattern Analysis and Classification with the New ACDA Seismic Signature Data Base, Report AD-A015925, DTIC, Alexandria, Va., 1975.
17. Anderson, K. R., Automatic Processing of Local Earthquake Data, Ph.D. thesis, Massachusetts Institute of Technology, Cambridge, 1978.
18. Chen, C. H., Ed., *Computer-Aided Seismic Analysis and Discrimination*, Elsevier, Amsterdam, 1978.
19. Bolt, B. A., *Nuclear Explosions and Earthquakes, The Parted Veil*, W. H. Freeman, San Francisco, 1976.
20. Landers, T. E. and Lacoss, R. T., Some geophysical applications of autoregressive spectral estimates, *IEEE Trans. Geosci. Electron.*, GE-15(1); 26, 1977.
21. Griffiths, L. J. and Prieto-Diaz, R., Spectral analysis of natural seismic events using autoregressive techniques, *IEEE Trans. on Geosci. Electron.*, GE-15(1), 13, 1977.

22. **Mendel, J. M.,** White-noise estimators for seismic data processing in oil exploration, *IEEE Trans. Autom. Control,* AC-22(5), 694, 1977.
23. **Silvia, M. T. and Robinson, E. A.,** *Deconvolution of Geophysical Time Series in the Exploration for Oil and Natural Gas,* Elsevier, Amsterdam, 1979.
24. **Claerbout, J. F.,** *Fundamentals of Geophysical Data Processing, with Applications to Petroleum Prospecting,* McGraw-Hill, New York, 1976.

Chapter 8

APPLICATIONS TO GEOPHYSICS: PART II

C. H. Chen

TABLE OF CONTENTS

I. INTRODUCTION

In this chapter two different geophysical signal processing application areas are discussed. The first area is adaptive digital filtering and related techniques for intrusion detection using seismic sensors. For this application a real-time signal processing capability is greatly needed. The second area is the high-resolution, nonlinear maximum entropy spectral analysis for a small number of geophysical data. The number of data points may be small because the data are not easily available, some data points are missing, or only a small segment of waveform is analyzed. The nonlinear maximum entropy method, though requiring considerably more computation, is far superior to Burg's maximum entropy method (see Chapter 3) in spectral resolution while eliminating the undesired spectrum-splitting phenomenon. The number of data points considered is typically less than 200.

II. ADAPTIVE DIGITAL FILTERING FOR INTRUSION DETECTION

Adaptive digital filtering is an effective method to remove or reduce noisy signals generated by sources at a large distance away from the seismic sensors that are used for intrusion detection. The noisy signals need to be removed because only those signals generated near the sensors are of interest for detection. The signals to be detected may correspond to the footsteps of a walking person. An earlier intensive study on this subject was reported by scientists at Sandia Laboratories.[1-5] The practical approach considered here is to use two or more spatially separated sensors. Each sensor may consist of a number of transducer elements, each contributing constructively to the total response of the sensor. Figure 1 is a sketch of the sensor layout. In this sketch, the seismic energy from the far-field source is very similar at both Sensors 1 and 2. Therefore, we should be able to reduce the final signal used for processing by removing the correlated portions of the signal observed at Sensors 1 and 2. When a person is very near to Sensor 1 (Sensor 2 being 3 m away or more), he generates a much different signal in Sensor 1 than in Sensor 2. Therefore, the signals of interest should not be correlated. Mathematically, the problem is to detect the footstep signal from the correlated background noise caused by, e.g., moving vehicles at some distance away plus additive sensor noises. The adaptive procedure is needed because of the nonstationary nature of the data considered.

Figure 2 is a typical set of four channel data along with their Fourier amplitude spectra. The fourth channel is a timer that marks the occurrence of the footstep signal or event. Each channel has 960 data points at a sampling rate of 200 samples per second. This data set has a strong footstep signal to background noise ratio. The following digital filter implementation is considered. Other adaptive digital filtering algorithms are available.[1,5]

The block diagram of the adaptive digital filter considered is shown in Figure 3A. The input-output relation of this nonrecursive adaptive digital filter, as shown in Figure 3B, is given by

$$g_m = \sum_{n=0}^{N} b_{n,m} f_{m-n} \qquad (1)$$

where g_m is the filter output, $b_{n,m}$ is the filter coefficient, and f_n is the filter output. Define the vectors

$$B_m' = [b_{0,m} \ b_{1,m} \quad \text{----} \quad b_{N,m}]$$

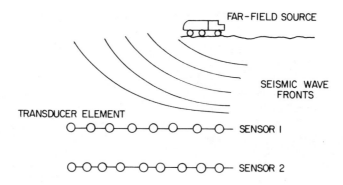

FIGURE 1. Sketch of seismic sensor layout for intrusion detection.

and

$$F_m' = [f_m \quad f_{m-1} \quad \cdots \quad f_{m-N}]$$

where N is the order of the filter and m is the number of iterations in adjusting the filter coefficients. Equation 1 can be written as

$$g_m = B_m'F_m = F_m'B_m \tag{2}$$

We use the Noisy Least-Mean-Square Algorithm[1] described in our computer program to calculate the filter coefficients $b_{n,m}$, i.e.,

$$B_{m+1} = B_m + v\, e_m F_m$$

$$e_m = y_m - g_m$$

$$g_m = B_m'F_m \tag{3}$$

where v is a constant. After adjusting the values of v, N, m, and delay time, we can obtain the desired result with a greatly improved signal-to-noise ratio. A detailed computer program listing is given in Appendix A. The meanings of some variables and arrays in the subroutine AF (for adaptive filtering) are stated below:

NDELAY: Number of delay points
M: Number of iterations
N: The order of the adaptive digital filter
V: The constant v of the adaptive digital filter

Figure 4 shows a sequence of filtered waveforms for Channel 1. The best results is given by Figure 4E. Here M = 1, NDELAY = 30, N = 18, and V = 0.0112. Figure 5 shows a sequence of filtered waveforms for Channel 2, with the best result given by Figure 5E. Here M = 1, NDELAY = 30, N = 17 and V = 0.00145. It is noted that the parameter choice is very sensitive in the adaptive digital filtering operations. The filtered results clearly illustrate the suppression of background noise and the enhancement of footstep signals.

The adaptive digital filtering described above is suitable for real-time processing. Kalman filtering is another important method with real-time processing capability. The

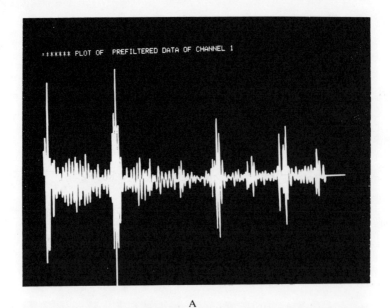

A

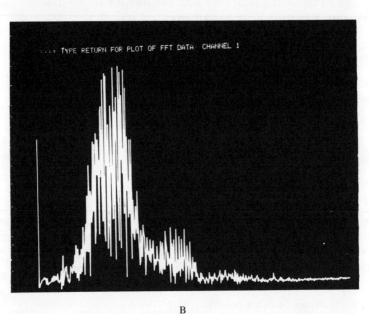

B

FIGURE 2.   Typical set of four channel data: Channel 1; (A), and its Fourier spectrum, (B); Channel 2, (C), and its Fourier spectrum, (D); Channel 3, (E), and its Fourier spectrum, (F); and Channel 4, (G), and its Fourier spectrum, (H).

signal model of the Kalman filter[6] is

$$x_{k+1} = F_k x_k + G_k w_k$$

$$z_k = y_k + v_k = H_k' x_k + v_k \tag{4}$$

where x_o, $\{v_k\}$, $\{w_k\}$ are jointly Gaussian and mutually independent; x_o is Gaussian distributed with mean $\bar{x}_o$ and covariance P_o, respectively; $\{v_k\}$ is zero mean, covariance

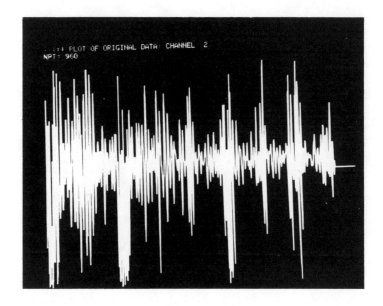

FIGURE 2C

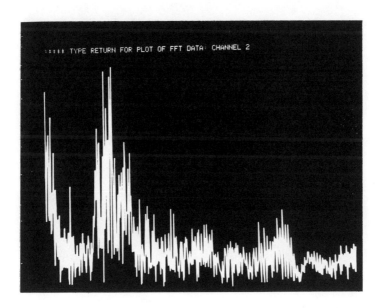

FIGURE 2D

$R_k \delta_k \ell$; and $\{w_k\}$ is zero mean, covariance $Q_k \delta_k \ell$. We can estimate x_k by the following formula:

$$\hat{x}_{k+1/k} = (F_k - K_k H_k') \hat{x}_{k/k-1} + K_k Z_k; \ \hat{x}_{0/-1} = \bar{x}_0$$

$$K_k = F_k \sum_{k/k-1} H_k'(H_k' \sum_{k/k-1} H_k + R_K)^{-1}$$

$$\sum_{k+1/k} = F_k \left[\sum_{k/k-1} - \sum_{k/k-1} H_k(H_k' \sum_{k/k-1} H_k + R_k)^{-1} \right.$$

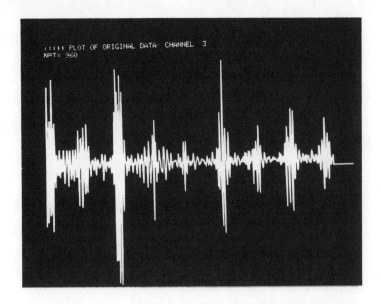

FIGURE 2E

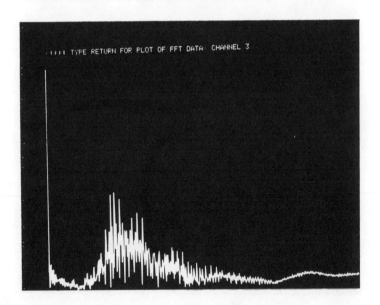

FIGURE 2F

$$\cdot H_k' \sum\nolimits_{k/k-1}\Big] F_k' + G_k Q_k G_k'; \; \sum\nolimits_{0/-1} = P_0$$

$$\hat{x}_{k/k} = \hat{x}_{k/k-1} + \sum\nolimits_{k/k-1} H_k H_k' \sum\nolimits_{k/k-1} H_k + R_k)^{-1}$$

$$\cdot (Z_k - H_k' \hat{x}_{k/k-1})$$

$$\sum\nolimits_{k/k} = \sum\nolimits_{k/k-1} - \sum\nolimits_{k/k-1} H_k (H_k' \sum\nolimits_{k/k-1} H_k + R_k)^{-1} H_k' \sum\nolimits_{k/k-1}$$

where $\hat{x}_{k/k}$ is the conditional estimate of x_k.

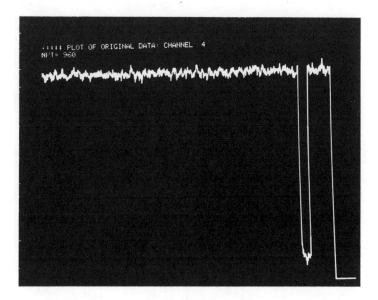

FIGURE 2G

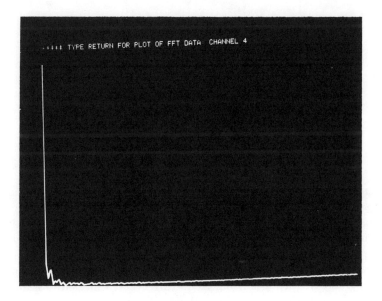

FIGURE 2H

In the computer study, G_k is fixed but F_k is estimated recursively for each k by using Equation 4. Figure 6 is the result of Kalman filtering by plotting $x_{k/k}$ for Channels 1, 2 and 3 with five iterations at each data point. For this set of data, the signal extraction capability of Kalman filtering appears to be slightly less than that of adaptive digital filtering, while the two methods require nearly the same amount of computation.

III. A NEW MAXIMUM ENTROPY METHOD FOR SPECTRAL ANALYSIS

In many geophysical problems the data records are short, while an accurate spectral

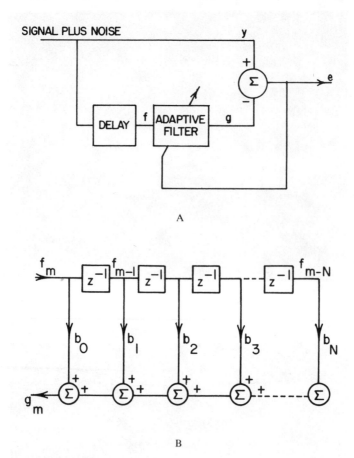

SIGNAL PLUS NOISE

FIGURE 3.   (A) Block diagram of the system using adaptive digital filter; (B) Input-output relation of the adaptive digital filter.

analysis is required. The conventional power spectrum estimation using smoothing and windowing procedures may not provide sufficient resolution, while causing undesired Gibbs phenomenon. Burg's maximum entropy method for spectrum analysis considerably improves the spectral resolution for short records. A discussion of Burg's method is given in Chapter 3. The method makes no assumption of data outside the time interval specified and is thus least committal on unavailable data. However, spectrum splitting and frequency shifting frequently occur, especially for sinusoidal signals.[7,8] Generally speaking, the method enhances the peaky components of the spectrum. A new maximum entropy power spectrum analysis method was proposed[9,10] that makes use of nonlinear optimization procedure and provides the lowest possible prediction error power under the stability constraints.

The following is a brief description of mathematical procedures with the new method.[9,10] Complex signals are considered, as the real signal is a special case. Given an n-point data sample $(x_1, x_2, ---, x_n)$ of complex numbers x_i, which may be formed by real data from two different channels. Define an $(m + 1)$ point prediction error filter (PEF) $(1, g_{m1}, g_{m2}, ---, g_{mm})$, where each g_{ij} is a complex variable such that the kth prediction errors are

$$e_{1k} = \sum_{i=0}^{m} x_{k+m-i} \, g_{mi}$$

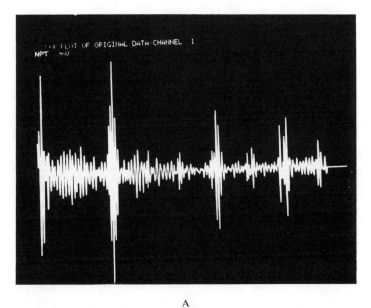

A

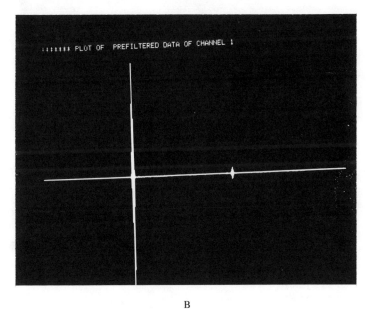

B

FIGURE 4. The sequence of filtered waveform for Channel 1; (A), at different parameter sets; (B—E).

$$e_{2k} = \sum_{i=0}^{m} x_{k+i}\, g_{mi}^{*}$$

$$k = 1, 2, 3, ---, i-m \tag{5}$$

where g_{mi}^{*} is the complex conjugate of g_{mi}, $g_{mo} = 1$, and e_{1k} and e_{2k} are the forward and backward prediction errors, respectively.

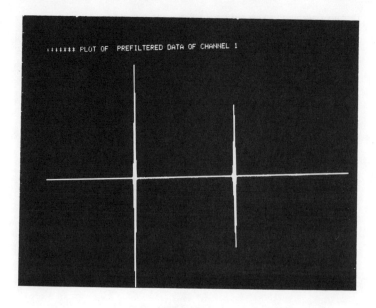

FIGURE 4C

Now, the mean square prediction error, or mean error power, in both time directions is

$$P_m = 0.5 \, (n-m)^{-1} \sum_{s=1}^{2} \sum_{k=1}^{n-m} e_{sk} \, e_{sk}{}^* \qquad (6)$$

If the PEFs (with leading "1" suppressed) of all orders 1, 2, ---, m are gathered in one complex matrix G_m, we may write

$$G_m = \begin{bmatrix} g_{11} & & & \\ g_{21} & g_{22} & & \\ \cdot & & & \\ \cdot & & & \\ \cdot & & & \\ g_{m1} & g_{m2} & --- & g_{mm} \end{bmatrix} \qquad (7)$$

The generalization of the Levinson recursion algorithm is given by

$$g_{jk} = g_{j-1,k} + g_{jj} \, g_{j-1,j-k}^* \qquad (8)$$

Burg has shown that if these diagonal elements (also called reflection coefficients) all lie in the range $|g_{jj}| < 1$, then the PEF is minimum phase, that is, its Z-transform has all its zeros outside the unit circle. It is noted here that the Z-transform defined by geophysicists differs from that defined by electrical engineers in the sign of complex exponents (see Chapter 2).

In order to enforce this condition we can set

$$g_{jj} = U \tanh\theta_j \, \exp i\phi_j \qquad (9)$$

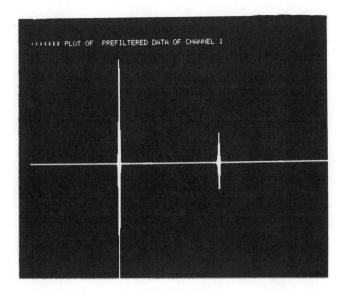

FIGURE 4D

for complex signals, where $0 \le \theta_j \le \pi$, θ_j and ϕ_j are both real and U is a positive constant slightly less than unity. U is adjusted so that all Z-transform roots of the PEF lie outside the unit circle and none lie on it. For real signals, we can set

$$g_{jj} = U\sin\theta_j \qquad (10)$$

The nonlinear optimization procedure starts with the expression for variations in θ_j, ϕ_j

$$\theta_j = \theta_j{}^0 + \Delta\theta_j$$
$$\phi_j = \phi_j{}^0 + \Delta\phi_j \qquad (11)$$

and then expand the prediction error e_{sk} in a Taylor series about $\theta_j{}^0$ and $\phi_j{}^0$, retaining only the first order terms. Then, set $\partial P_m/\partial \Delta\theta_j = 0$, $\partial P_m/\partial \Delta\phi_j = 0$ to find the corrections $\Delta\theta_j$ and $\Delta\phi_j$. The corrections are then substituted into Equation 11 and the process is repeated until the corrections $\Delta\theta_j$ and $\Delta\phi_j$ become sufficiently small. The method of Fletcher and Powell[11] is utilized in the nonlinear iterative procedure for error minimization. In the case of complex signals, error minimization procedure by the method of conjugate gradient[12] is also utilized. Detailed computer listings of the new (nonlinear) maximum entropy spectral analysis algorithm are presented in Appendix B.

The new maximum entropy power spectral analysis method was tested by the author for both real and complex signals. For real signals of various levels of additive Gaussian noise, the cases considered include pure sinewave of various initial phases, the sum of two sinewaves of different frequencies, and the FSK (frequency shift keying) signal consisting of two sinewaves of different frequencies occupying nonoverlapped time intervals. The nonlinear method consistently corrects the line-splitting and frequency shifting. Line-splitting is most evident when the initial phase of the sinewave is an odd multiple of 45°. Figure 7 shows the power spectra of a 1 Hz 16-point sinewave with initial phase of 45° and 6 filter weights. The nonlinear method of Fougere clearly corrects the line-splitting in Burg's method, and the line spectrum at 1 Hz is consider-

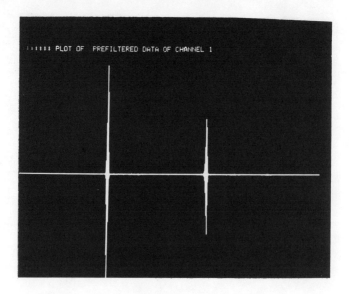

FIGURE 4E

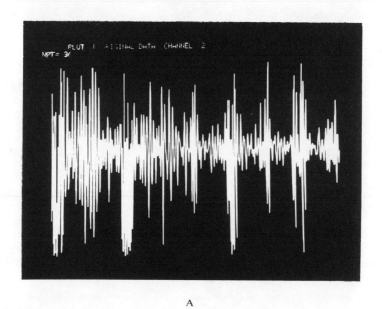

A

FIGURE 5.   The sequence of filtered waveform for Channel 2; (A), at different
parameter sets; (B—E).

ably sharper. A logarithmic plot of the power spectrum density is also provided. The
additive noise level has a noise power of 0.0002. As the noise power increases, the line-
splitting phenomenon gradually disappears, but frequency shifting becomes more ap-
parent. Again, the nonlinear method is superior, as it corrects this problem also. For
multiple equal-amplitude sinewaves the nonlinear method always improves the resolution, even though equal-amplitude spectral peaks cannot be guaranteed. For the FSK
signal, the computer results show that the nonlinear method always provides correct
locations of spectral peaks while Burg's method often shows frequency shifting. The

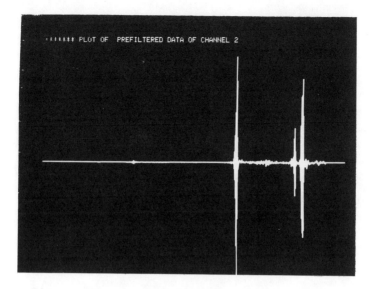

FIGURE 5B

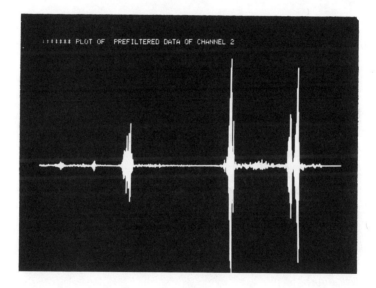

FIGURE 5C

algorithm for real signals by using the nonlinear method has been fully developed. It includes Burg's method and has been applied to geophysical data such as geomagnetic micropulsations[13] and sunspots.[14]

For the complex signals, a 16-point complex sinusoid and one data set provided by RADC for 1979 Spectral Estimation Workshop are considered. For the complex sinusoid, spectral splitting is not evident. However, the nonlinear method provides much sharper spectral peaks. Figure 8 shows the two channels (I channel and Q channel) of RADC data set with 32 points in each channel. A complex signal consists of the I and Q channels of data as real and imaginary parts, respectively. The given data set contains a number of radar targets, each of which is a complex sinusoid at a single frequency, in a cluttered environment. Each target has been sampled at least twice per cycle, but no more than six times per cycle. A target may have positive or negative

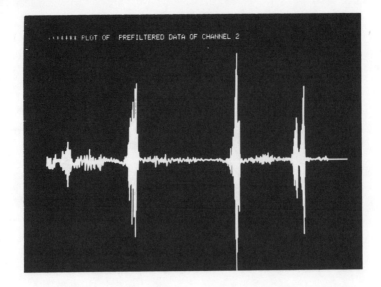

FIGURE 5D

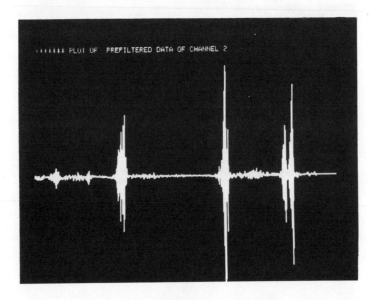

FIGURE 5E

frequency. The clutter is a random process with spectrum centered at zero frequency. The clutter spectrum follows the expression $\exp(-(8R)^2)$ for $-1/6 < R < 1/6$, where R is the ratio of observation frequency to the sampling rate f_s. The clutter spectrum is known to be zero within the range of possible target frequencies.

As a result of the limited time during which data was collected, a discrete Fourier transform on the given data will not completely resolve the targets or separate them entirely from the clutter process. Spectrum estimation techniques must be used to locate the target frequencies and determine their amplitudes. The true answer[15] for frequencies, amplitudes, and relative levels in dB is given as follows:

Frequency	Amplitude	Relative level
$+0.1836f_s$	0.2184	0 dB

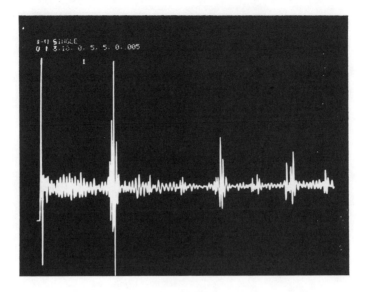

A

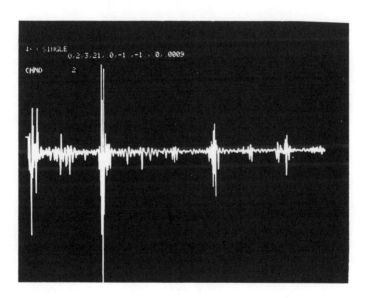

B

FIGURE 6.   Result of Kalman filtering with five iterations for; (A) Channel 1,
(B) Channel 2, and (C) Channel 3.

$+0.2930f_s$	0.0873	-7.9dB
$-0.3945f_s$	0.1203	-5.18dB

Here, the positive frequency corresponds to a target that is receding from the observer.

In the computer study, the power spectra for different filter weights are considered. By using three filter weights, neither Burg's method nor the nonlinear method is adequate to provide enough spectral resolution. For ten filter weights, there are sufficient details in the spectrum, as shown by Figure 9. For 16 filter weights, the spectral content

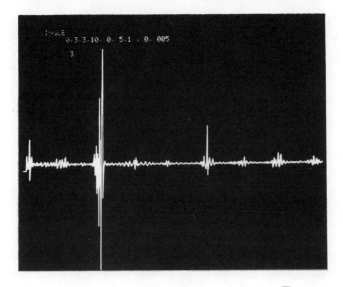

FIGURE 6C

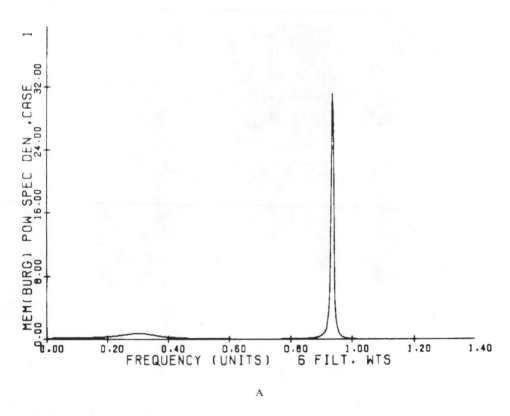

A

FIGURE 7.   Power spectra of a 1 Hz 16-point pure sinewave with initial phase of 45° and 6 filtered weights. (A) Burg's method power spectrum, (B) logarithmic plot of (A), (C) nonlinear method power spectrum, and (D) logarithmic plot of (C).

is about the same and there is no improvement in resolution. In both 10 and 16 filter weight cases, the nonlinear method is superior to the Burg's method. Figure 9 identifies correctly the three target frequencies and their relative amplitude levels.

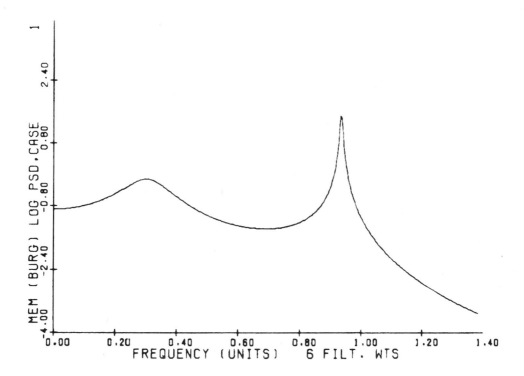

FIGURE 7B

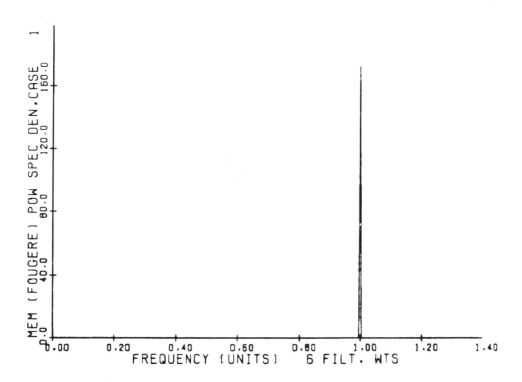

FIGURE 7C

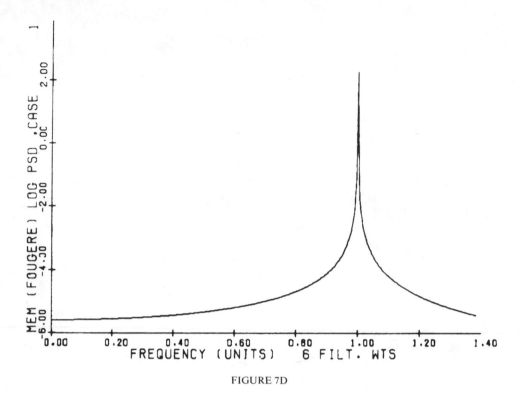

FIGURE 7D

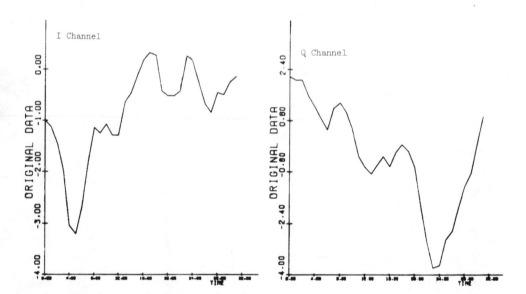

FIGURE 8.   An RADC, two-channel data set that forms a 32-point complex signal.

The solid curve in Figure 10 is the final prediction error (FPE) for the RADC data set described above, based on the Akaike criterion,[16] defined as

$$(\text{FPE})_m = \frac{n + (m+1)}{n - (m+1)} \; P_m \tag{12}$$

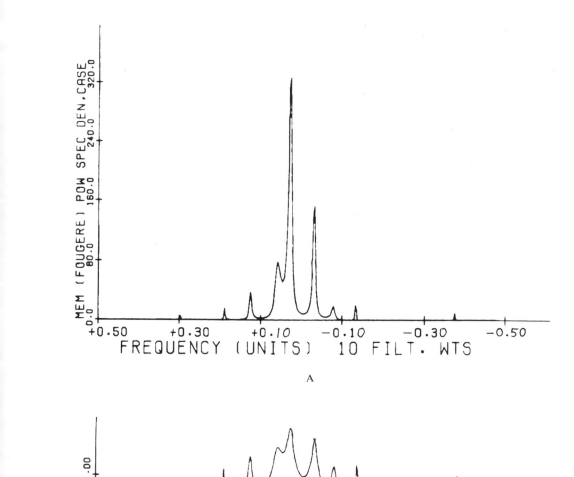

A

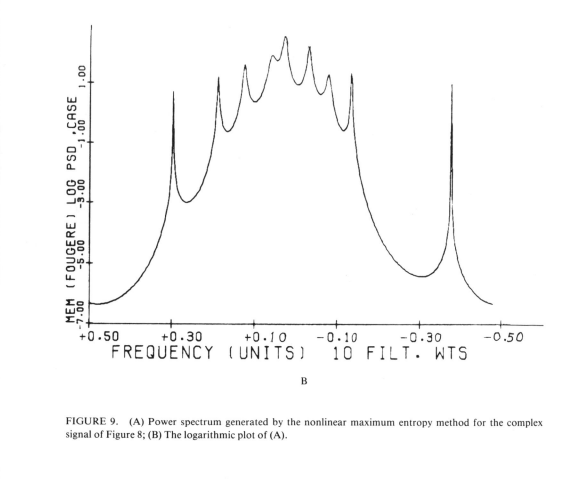

B

FIGURE 9. (A) Power spectrum generated by the nonlinear maximum entropy method for the complex signal of Figure 8; (B) The logarithmic plot of (A).

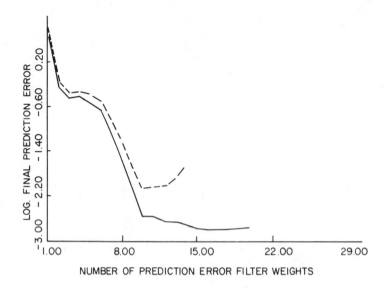

FIGURE 10.   Plot of the final prediction error as a function of the number of
PEF weights for the data set of Figure 8.

where n is the number of data points in one channel, and m is the number of prediction
error filter weights or the order of the filter. The Akaike criterion takes into account
only the forward prediction error, while the maximum entropy method considers both
the forward and backward prediction errors. Because of this, a possible modification
of Equation 12 is

$$(FPE)_m = \frac{n + 2(m + 1)}{n - 2(m + 1)} \, P_m \qquad (13)$$

which is plotted for the same data set as dashed curve in Figure 10. Equation 12 suggests a minimum final prediction error at m = 16, while Equation 13 suggests the
minimum at m = 10. It appears that the true final prediction error has Equations 12
and 13 as lower and upper bounds, respectively, and the error is closer to the upper
bound.

As a bibliographic note, the Proceedings of the May 1978 and October 1979 RADC
Spectrum Estimation Workshops have a number of important articles on maximum
entropy and related spectral estimation techniques. Two edited books[17,18] and a bibliography[19] are additional sources.

Appendix A

PROGRAM SPAN4 FOR ADAPTIVE DIGITAL FILTERING OF
SEISMIC DATA

```
DIMENSION F (4, 32), X(1024), C(257)
COMPLEX CY (1024)
INTEGER CHNO, PREF, DLY, ITR, ODR
CUMMUN F, NF, INDEX
DEFINE FILE 3 (30, 256, U, INDEX)
```

```
6      FORMAT ('0')
5      FORMAT (' INPUT IGD, NP, PREF, DLY, ITR, ODR, NTE, V')
7      FORMAT (715, F12 5)
       WRITE (6, 5)
       WRITE (6, 6)
       READ (6, 7) IGD, NP, PREF, DLY, ITR, ODR, NTE, V
       IF (IGD, NF, 0) GO TO 71
       CALL GDATA

71     NF = 960
       DO 8 I = 1, 1024
8      X (I) = FLOAT (I)
       DO 10 CHNO = 1, 4
       INDEX = 1
       DO 112 J = 1, 30
       READ (3'INDEX)F
       L = (J-1)*32
       DO 15 I = 1, 32
       L = L + 1
       Y (L) = F (CHNO, I)
15     CONTINUE
112    CONTINUE
       DO 352 I = 1, NF
       Y (I) = Y (I)/1000
352    CONTINUE
       DO 419 I = 961, 1024
       Y (I) = 0
419    CONTINUE
20     FORMAT ('***** PLOT OF ORIGINAL DATAANNEL', I3)
       IF (NP. NE. 0) GO TO 56
       WRITE (5,54) CHNO
54     FORMAT ('1******** ORIGINAL DATA FUNCTION VALUES OF
       CHANNEL', I2)
       WRITE (5,55) Y
55     FORMAT (1X,10F13 5)
56     WRITE (6, 21) NF
21     FORMAT (' NPT = ', I4)
       READ (6, 30) NX
       CALL NFWPAG
       WRITE (6, 20) CHNO
       WRITE (6, 21) NF
30     FORMAT (I3)
       CALL KBPLOT (X, Y, 1024, 0, 1023, 10, 700)
       READ (6, 30) NX
       CALL NEWPAG
       IF (PREP, NE, 1) GO TO 931
       CALL AF (Y, NF, DLY, ITR, ODR, NIE, V)

931    DO 939 I = 1, 1024
       CY (I) = CMPLX (Y(I), 0.0)
939    CONTINUE
```

```
        IF (PREP, EQ. 0) GO TO 2100
        WRITE (6, 87) CHNO
87      FORMAT (' 1******* PLOT OF PREFILTERED DATA OF CHANNEL',
        I2)
        CALL KBPLOT (X, Y, 1024, 1, 1000, 10, 700)
        READ (6,30) NX
        CALL NEWPAG
        IF (NP, NE, O) GO TO 2100
        WRITE (5,2110) CHNO
2110    FORMAT ('1 ******* PREFILTERED DATA OF CHANNEL', I2)
        WRITE (5, 55) Y
2100    CALL ANFFT (CY, 10)
        DO 950 I = 1, 1024
        Y (I) = CABS (CY(I))
950     CONTINUE
        WRITE (6, 50) CHNO
50      FORMAT (' ***** TYPE RERUN FOR PLOT OF FFT DATA: CHANNEL
        ', I2)
        READ (6, 30) NX
        CALL KBPLOT(X, Y, 512, 0, 1023, 10, 700)
        READ (6,30) NX
        IF (NP. NE. 0) GO TO 175
        WRITE (5, 171) CHNO
171     FORMAT ('1******** FFT DATA FUNCTION VALUES OF CHANNEL',
        I2)
        WRITE (5, 55)Y
175     READ (6, 30) NX
10      CONTINUE
        CALL EXIT
        END
        SUBROUTINE GDATA
        INTEGER CH, RATE
        BYTE B (1280)
        DIMENSION IA (1280), F (4, 32)
        COMMON F, NF, INDEX
        INDEX = 1
        DO 200 NR = 1,30
        CALL READUN (B, NPT)
        NF = NPT/8*6
        WRITE (6, 50)NPT
50      FORMAT ('0 TOTAL BYTES READ FROM TAPE = ', I7)
        DO 70 I = 1, NPT
        Call BY2IN (B(I), II)

        IA (I) = II
70      CONTINUE
85      FORMAT (12I5)
        DO 80 I = 1, NPT, 2
        IA (I + 1)/2) = IA(I)*256 + IA(I + 1)
80      CONTINUE
        NT = NPT/2
        DO 90 CH = 1, 4
```

```
         K = 1
         DO 90 I = CH, NT, 20
         F (CH,K) = FLOAT (IA(I))
90       K = K + 1
         WRITE (3'INDEX) F
200      CONTINUE
         RETURN
         END
         SUBROUTINE ANFFT (X,M)
         COMPLEX X (1), U,W,T
         N = 2**M
         N2 = N/2
         N1 = N-1
         J = 1
         DO 3 I = 1, N1
         IF (I. GE J) GO TO 1
         T = X (J)
         X (J) = X (I)
         X (I) = T
1        K = N2
2        IF (K, GE, J) GO TO 3
         J = J-K
         K = K/2
         GO TO 2
3        J = J + K
         PI = 3. 1415926
         DO 5 L = 1, M
         LE = 2**L
         LE1 = LE/2
         U = (1. 0, 0. 0)
         W = CMPLX(COS (PI/LE1), SIN(PI/LE1))
         DO 5 J = 1, LE1
         DO 4 I = J, N, LE
         ID = I + LE1
         T = X(ID)*U
         X(ID) = X(I)-T
4        X (I) = X (I) + T
5        U = U*W
         RETURN
         END
         SUBROUTINE AF (ND, NF, NDELAY, M, N, NTE, V)
         REAL ND(1), F(40), B(40), DND(400)
         DOUBLE PRECISION G
         G = 0
         NP = N + 1
         DO 50 I = 1, NP
         B(I) = 0
         F(I) = 0
50       CONTINUE
         DO 40 I = 1, NDELAY
         J = NDELAY-I + 1
```

```
          DND(J) = ND(I)
   40     CONTINUE
          IIP = NDELAY + 1
          DO 100 IND = IIP, NF
          Y = ND(IND)
          CALL SHIFTR (Y, DND, NDELAY, TEMP)
          CALL SHIFTR (TEMP, F, NP, TEMP1)
          DO 200 IM = 1, M
          G = 0
          DO 500 I = 1, NP
          G = G + B(I)*F(I)
   500    CONTINUE
          E = Y−G
          DO 400 I = 1, NP
          B(I) = B(I) + V*E*F(I)
   400    CONTINUE
          IF (NTE. NE. 0) GO TO 200
          PRINT 600, IND, IM, E, G
   600    FORMAT (1X, 'SAMPLE', I5, 'ITERATION', I5, 'E = ', F15, 5,'G = ', F15.
          5)
          PRINT 601, B
   601    FORMAT (1X, 12F10. 2)
          PRINT 602, F
   602    FORMAT (1X, 12F10. 2)
   200    CONTINUE
   700    ND (IND) = G
   100    CONTINUE
          RETURN
          END
          SUBROUTINE SHIFTR (SHIN, A, L, SOUT)
          REAL A(1)
          SOUT = A(L)
          NDIM = L−1
          DO 100 K = 1, NDIM
          I = L−K + 1
          J = I−1
          A(I) = A(J)
   100    CONTINUE
          A(1) = SHIN
          RETURN
          END
```

Appendix B

COMPUTER PROGRAM LISTING OF NONLINEAR MAXIMUM ENTROPY SPECTRAL ANALYSIS FOR COMPLEX SIGNALS

```
C
C
C       NAME: OPTM  [100,100]  MAX. ENTROPY PROB. DISK II

        DOUBLE PRECISION XR(25), XI(25), U, FM, POWER, EXTA(1048)
        REAL XB(251), CAL(251)
```

```
        COMMON /BLKO/LX,LA,U,PM,POWER/BLK1/XR,XI/BLK2/EXTA
        NF=251
        U=.99999999D0
        READ(6,10)F1,F2,A1R,A1I,A2R,A2I,SR,LX,LA,LG,ICA,LIMIT,SNR
10      FORMAT(7F12.5,/,5I5,F12.5)
        IF(LG.GT.0)WRITE(5,5)
5       FORMAT(1H1,//,1X,'        F1       F2      A1R      A1I      A2R      A2I
       1     SR       LG     ICA   LIMIT     SNR',//)
        IF(LG.GT.0)WRITE(5,6)F1,F2,A1R,A1I,A2R,A2I,SR,LG,ICA,LIMIT,SNR
6       FORMAT(/,1X,7F8.2,3I8,F8.2)
        DO 20 I=1,NF
20      XB(I)=FLOAT(I)
        RNMAG=GC(SNR)
        CALL GENC(F1,F2,45.,0.,A1R,A1I,A2R,A2I,SR,RNMAG)
        DO 30 I=1,LX
30      CAL(I)=XR(I)
        CALL NEWPAG
        CALL KBPLOT(XB,CAL,LX,10,1000,400,700)
        DO 40 I=1,LX
40      CAL(I)=XI(I)
        CALL KBPLOT(XB,CAL,LX,10,1000,50,350)
        CALL BELL
        READ(6,50)NX
50      FORMAT(10I10)
        CALL BURG(ICA)
        IF(LG.GE.0)CALL MESG
        IF(ICA.EQ.0)GO TO 60
        CALL OPTM(LIMIT,LG)
        CALL FGR(1)
        IF(LG.GE.0)CALL MESG
60      CALL SPEC(SR,NF)
        KK=0
65      DO 70 I=1,NF
70      CAL(I)=EXTA(454+I)
        CALL KBPLOT(XB,CAL,NF,10,1000,100,700)
        DO 80 I=1,NF
        CAL(I)=0.
80      IF(MOD(I,25).EQ.1)CAL(I)=1.
        CAL(1)=50.
        CALL KBPLOT(XB,CAL,NF,10,1000,100,700)
        IF(KK.NE.0)GO TO 766
        IF(LG.EQ.-1)GO TO 100
        WRITE(5,336)
336     FORMAT(1X,///,1X,' THE SPECTRUM MAG ',//,'     HZ      MAG.',//)
        FREQ1=SR/(2.*FLOAT(NF-1))
        XB(1)=-SR/4.
        CAL(1)=EXTA(454+1)
        DO 199 I=2,NF
        CAL(I)=EXTA(454+I)
199     XB(I)=XB(I-1)+FREQ1
        WRITE(5,371)(XB(I),CAL(I),I=1,NF)
371     FORMAT(1X,5(F10.5,2X,E11.4,1H;),/)
100     DO 90 LL=1,NF
        IF(EXTA(454+LL).GT.0.)EXTA(454+LL)=ALOG10(EXTA(454+LL))
90      CONTINUE
        KK=KK+1
        CALL BELL
        READ(6,50)NX
        GO TO 65
766     CALL BELL
        CALL BELL
        CALL EXIT
        END


        FUNCTION GC(SNR)
        VA=SNR/10.
        VB=1./(2.*(10.**VA))
        GC=SQRT(VB)
        RETURN
        END
```

```
      SUBROUTINE GENC(F1,F2,P1,P2,A1R,A1I,A2R,A2I,SR,RNMAG)
      DOUBLE PRECISION XR(25),XI(25),U,PM,POWER
      COMMON /BLKO/LX,LA,U,PM,POWER/BLK1/XR,XI
      SP=1./SR
      PI=ATAN(1.)*4.
      W1=2.*PI*F1
      W2=2.*PI*F2
      P1=P1/180.*PI
      P2=P2/180.*PI
      POWER=0.D0
      T=0.
      I1=1
      I2=2
      DO 10 I=1,LX
      XR(I)=A1R*COS(W1*T+P1)+A2R*COS(W2*T+P2)+RNMAG*GAUSS(I1,I2)
      XI(I)=A1I*SIN(W1*T+P1)+A2I*SIN(W2*T+P2)+RNMAG*GAUSS(I1,I2)
      POWER=POWER+(XR(I)**2+XI(I)**2)
10    T=T+SP
      POWER=POWER/DFLOAT(LX)
      RETURN
      END


      FUNCTION GAUSS(I1,I2)
      GAUSS=0.
      DO 10 I=1,48
10    GAUSS=GAUSS+RAN(I1,I2)
      GAUSS=(GAUSS-24.)/2.
      RETURN
      END


      SUBROUTINE BURG(ICA)
      DIMENSION H(12),PEF(25),PER(25),F(25),G(12),GGG(12)
      COMPLEX G,PEF,PER,H,GGG,F,P,Q,SN,SD
      DOUBLE PRECISION XR(25),XI(25),AR(12),AI(12),GR(12,12)
     1,GI(12,12),U,PM,POWER,THETA(30),TMPR,TMP,EXTA(1048)
      COMMON /BLKO/NUM,LA,U,PM,POWER/BLK1/XR,XI/BLK2/EXTA
      EQUIVALENCE (EXTA(1),GR(1,1)),(EXTA(145),GI(1,1)),
     1(EXTA(289),AR(1)),(EXTA(301),AI(1)),(EXTA(313),THETA(1))
     2,(EXTA(343),PEF(1)),(EXTA(368),PER(1)),(EXTA(393),F(1)),
     3(EXTA(418),H(1)),(EXTA(430),G(1)),(EXTA(442),GGG(1))
      PM=POWER
      DO 10 I=1,NUM
10    F(I)=CMPLX(SNGL(XR(I)),SNGL(XI(I)))
      DO 110 NN=1,LA
      XERO=(0.,0.)
      N=NN-1
      IF(N.NE.0)GO TO 30
      DO 20 J=1,NUM
      PEF(J)=XERO
20    PER(J)=XERO
30    SN=XERO
      SD=XERO
      JJ=NUM-N-1
      DO 40 J=1,JJ
      Q=F(J+N+1)+PEF(J)
      P=F(J)+PER(J)
      SN=SN+CONJG(P)*Q
      SD=SD+P*CONJG(P)+Q*CONJG(Q)
40    CONTINUE
      GGG(NN)=-2.*SN/SD
      GR(NN,NN)=REAL(GGG(NN))
      GI(NN,NN)=AIMAG(GGG(NN))
      TMPR=CABS(GGG(NN))
      IF(TMPR.GE.1)TMPR=.99999
      IF(ICA.EQ.0)GO TO 50
      TMP=TMPR/U
      THETA(NN)=DATAN(TMP/DSQRT(1.D0-TMP*TMP))
      THETA(NN+LA)=DATAN2(GI(NN,NN),GR(NN,NN))
50    PM=PM*(1.-TMPR*TMPR)
```

```
            IF(N.EQ.0)GO TO 80
            DO 60 J=1,N
            K=N-J+1
            H(J)=GGG(J)+GGG(NN)*CONJG(GGG(K))
60          CONTINUE
            DO 70 J=1,N
70          GGG(J)=H(J)
            JJ=JJ-1
80          DO 90 J=1,JJ
            PER(J)=PER(J)+CONJG(GGG(NN))*(PEF(J)+F(J+NN))
            PEF(J)=PEF(J+1)+GGG(NN)*(PER(J+1)+F(J+1))
90          CONTINUE
            DO 100 I=1,NN
100         G(I)=GGG(I)
110         CONTINUE
            DO 120 I=1,LA
            AR(I)=REAL(G(I))
120         AI(I)=AIMAG(G(I))
            RETURN
            END



            SUBROUTINE SPEC(SR,NF)
            DOUBLE PRECISION AR(12),AI(12),PM,DT,ANG,U,EXTS,TPI,EXTA(1048)
           1,F,FINC,SCSR,SCSI
            COMMON /BLK0/LX,M,U,PM,EXTS/BLK2/EXTA
            EQUIVALENCE (EXTA(289),AR(1)),(EXTA(301),AI(1))
            TPI=8.D0*DATAN(1.D0)
            DT=1./SR
            F=-SR/4.
            FINC=SR/((NF-1)*2.)
            DO 5 I=1,NF
            SCSR=1.D0
            SCSI=0.D0
            DO 10 K=1,M
            ANG=-TPI*K*F*DT
            SCSR=SCSR+AR(K)*DCOS(ANG)-AI(K)*DSIN(ANG)
10          SCSI=SCSI+AR(K)*DSIN(ANG)+AI(K)*DCOS(ANG)
            EXTA(454+I)=PM*DT/(SCSR*SCSR+SCSI*SCSI)
5           F=F+FINC
            RETURN
            END



            SUBROUTINE FGR(ISW)
            DOUBLE PRECISION XR(25),XI(25),FR(25),FI(25),
           1BR(25),BI(25),THETA(30),AR(12),AI(12),
           2GR(12,12),GI(12,12),PS(30),PGPR(12,12),
           3PGPI(12,12),PGMR(12,12),PGMI(12,12),
           4U,PM,TMPR,TMPI,TMQR,TMQI,T1R,T1I,
           5T2R,T2I,P1R,P1I,P2R,P2I,POWER,EXTA(1048)
            COMMON /BLK0/N,M,U,PM,POWER/BLK1/XR,XI/BLK2/EXTA
            EQUIVALENCE (EXTA(1),GR(1,1)),(EXTA(145),GI(1,1)),
           + (EXTA(289),AR(1)),(EXTA(301),AI(1)),(EXTA(313),THETA(1)),
           + (EXTA(343),PS(1)),(EXTA(373),FR(1)),(EXTA(398),FI(1)),
           + (EXTA(423),BR(1)),(EXTA(448),BI(1)),
           + (EXTA(473),PGPR(1,1)),(EXTA(617),PGPI(1,1)),
           + (EXTA(761),PGMR(1,1)),(EXTA(905),PGMI(1,1))
            DO 310 J=1,M
            GR(J,J)=U*DSIN(THETA(J))*DCOS(THETA(M+J))
310         GI(J,J)=U*DSIN(THETA(J))*DSIN(THETA(M+J))
            DO 320 J=2,M
            J1=J-1
            DO 320 K=1,J1
            CALL CM(GR(J,J),GI(J,J),GR(J-1,J-K),-GI(J-1,J-K)
           + ,TMPR,TMPI)
320         CALL CAD(GR(J-1,K),GI(J-1,K),TMPR,TMPI,
           + GR(J,K),GI(J,K))
            IF(ISW.EQ.0)GO TO 377
            DO 379 J=1,M
            AR(J)=GR(M,J)
```

```
379       AI(J)=GI(M,J)
          RETURN
377       PM=0.DO
          NM=N-M
          DO 410 K=1,NM
          FR(K)=XR(K+M)
          FI(K)=XI(K+M)
          BR(K)=XR(K)
          BI(K)=XI(K)
          DO 420 I=1,M
          CALL CM(XR(K+M-I),XI(K+M-I),GR(M,I),GI(M,I)
      +   ,TMPR,TMPI)
          CALL CADE(FR(K),FI(K),TMPR,TMPI)
          CALL CM(XR(K+I),XI(K+I),GR(M,I),-GI(M,I)
      +   ,TMPR,TMPI)
420       CALL CADE(BR(K),BI(K),TMPR,TMPI)
410       PM=PM+FR(K)*FR(K)+FI(K)*FI(K)+BR(K)*BR(K)+
      +   BI(K)*BI(K)
          PM=PM/(2.DO*DFLOAT(NM))
          DO 5 J=1,M
          DO 510 I=1,M
          PGPR(I,I)=0.DO
          PGPI(I,I)=0.DO
          PGMR(I,I)=0.DO
          PGMI(I,I)=0.DO
          IF(I.EQ.J)PGPR(I,I)=1.DO
510       IF(I.EQ.J)PGMR(I,I)=1.DO
          DO 550 I=2,M
          I1=I-1
          DO 550 K=1,I1
          CALL CM(GR(I-1,I-K),-GI(I-1,I-K),PGPR(I,I),
      +   PGPI(I,I),TMPR,TMPI)
          CALL CM(GR(I,I),GI(I,I),PGPR(I-1,I-K),-PGPI(I-1,I-K)
      +   ,TMQR,TMQI)
          PGPR(I,K)=PGPR(I-1,K)+TMPR+TMQR
          PGPI(I,K)=PGPI(I-1,K)+TMPI+TMQI
          CALL CM(GR(I-1,I-K),-GI(I-1,I-K),
      +   PGMR(I,I),PGMI(I,I),TMPR,TMPI)
          CALL CM(GR(I,I),GI(I,I),
      +   -PGMR(I-1,I-K),PGMI(I-1,I-K),TMQR,TMQI)
          PGMR(I,K)=PGMR(I-1,K)+TMPR+TMQR
          PGMI(I,K)=PGMI(I-1,K)+TMPI+TMQI
550       CONTINUE
          TMPR=U*DCOS(THETA(J))*DCOS(THETA(J+M))
          TMPI=U*DCOS(THETA(J))*DSIN(THETA(J+M))
          TMQR=U*DSIN(THETA(J))*(-DSIN(THETA(J+M)))
          TMQI=U*DSIN(THETA(J))*DCOS(THETA(J+M))
          DO 25 I=1,M
          PGPR(1,I)=TMPR*PGPR(M,I)+TMPI*(-PGMI(M,I))
          PGPI(1,I)=TMPR*PGPI(M,I)+TMPI*PGMR(M,I)
          PGMR(1,I)=TMQR*PGPR(M,I)+TMQI*(-PGMI(M,I))
25        PGMI(1,I)=TMQR*PGPI(M,I)+TMQI*PGMR(M,I)
          PS(J)=0.DO
          PS(J+M)=0.DO
          DO 20 K=1,NM
          T1R=0.DO
          T1I=0.DO
          T2R=0.DO
          T2I=0.DO
          P1R=0.DO
          P1I=0.DO
          P2R=0.DO
          P2I=0.DO
          DO 30 I=1,M
          CALL CM(XR(K+M-I),-XI(K+M-I),
      +   PGPR(1,I),-PGPI(1,I),TMPR,TMPI)
          CALL CADE(T1R,T1I,TMPR,TMPI)
          CALL CM(XR(K+I),-XI(K+I),
      +   PGPR(1,I),PGPI(1,I),TMPR,TMPI)
          CALL CADE(T2R,T2I,TMPR,TMPI)
          CALL CM(XR(K+M-I),-XI(K+M-I),
```

```
     +       PGMR(1,I),-PGMI(1,I),TMPR,TMPI)
             CALL CADE(P1R,P1I,TMPR,TMPI)
             CALL CM(XR(K+I),-XI(K+I),PGMR(1,I),PGMI(1,I),TMPR,TMPI)
30           CALL CADE(P2R,P2I,TMPR,TMPI)
             CALL CM(FR(K),FI(K),T1R,T1I,TMPR,TMPI)
             CALL CM(BR(K),BI(K),T2R,T2I,TMQR,TMQI)
             PS(J)=PS(J)+TMPR+TMQR
             CALL CM(FR(K),FI(K),P1R,P1I,TMPR,TMPI)
             CALL CM(BR(K),BI(K),P2R,P2I,TMQR,TMQI)
20           PS(J+M)=PS(J+M)+TMPR+TMQR
             PS(J)=PS(J)/DFLOAT(NM)
5            PS(J+M)=PS(J+M)/DFLOAT(NM)
             RETURN
             END


             SUBROUTINE CM(XR,XI,YR,YI,ZR,ZI)
             DOUBLE PRECISION XR,XI,YR,YI,ZR,ZI
             ZR=XR*YR-XI*YI
             ZI=XR*YI+XI*YR
             RETURN
             END


             SUBROUTINE CAD(XR,XI,YR,YI,ZR,ZI)
             DOUBLE PRECISION XR,XI,YR,YI,ZR,ZI
             ZR=XR+YR
             ZI=XI+YI
             RETURN
             END


             SUBROUTINE CADE(XR,XI,YR,YI)
             DOUBLE PRECISION XR,XI,YR,YI
             XR=XR+YR
             XI=XI+YI
             RETURN
             END


             SUBROUTINE OPTM(LIMIT,LG)
             DOUBLE PRECISION H(150),X(30),GRATH(30)
     +       ,F,OLDF,GRASUM,T,SQF,DY,HNRM,GNRM,EST,EPS
     +       ,FY,ALFA,AMBDA,FX,DX,Z,DALFA,W,UFR,EXTA(1048),FXTS
             COMMON /BLK0/LX,N,UER,F,FXTS/BLK2/EXTA
             EQUIVALENCE (EXTA(313),X(1)),(EXTA(343),GRATH(1))
             LUN=5
             IF(LG.EQ.-1)LUN=6
             EPS=10.D0**(-16)
             WRITE(5,2222)
2222         FORMAT(1X,//,' OPTIMIZATION',//)
             WRITE(LUN,5)
5            FORMAT(' COUNT  RMS GRADIENT  RMS ERROR')
             CALL FGR(0)
             IER=0
             KOUNT=0
             N2=N+N
             N3=N2+N
             N31=N3+1
10           K=N31
             DO 40 J=1,N
             H(K)=1.D0
             NJ=N-J
             IF(NJ)50,50,20
20           DO 30 L=1,NJ
             KL=K+L
30           H(KL)=0.D0
40           K=KL+1
50           KOUNT=KOUNT+1
             OLDF=F
             GRASUM=0.D0
```

```
          DO 90 J=1,N
          GRASUM=GRASUM+GRATH(J)**2
          K=N+J
          H(K)=GRATH(J)
          K=K+N
          H(K)=X(J)
          K=J+N3
          T=0.DO
          DO 80 L=1,N
          T=T-GRATH(L)*H(K)
          IF(L-J)60,70,70
60        K=K+N-L
          GO TO 80
70        K=K+1
80        CONTINUE
90        H(J)=T
          GRASUM=DSQRT(GRASUM/DFLOAT(N))
          SQF=DSQRT(F)
          IF(MOD(KOUNT,1).EQ.0)WRITE(LUN,670)KOUNT,GRASUM,SQF
          DY=0.DO
          HNRM=0.DO
          GNRM=0.DO
          DO 100 J=1,N
          HNRM=HNRM+DABS(H(J))
          GNRM=GNRM+DABS(GRATH(J))
100       DY=DY+H(J)*GRATH(J)
          IF(DY)110,540,540
110       IF(HNRM/GNRM-EPS)540,540,120
120       FY=F
          ALFA=2.DO*(EST-F)/DY
          AMBDA=1.DO
          IF(ALFA)150,150,130
130       IF(ALFA-AMBDA)140,150,150
140       AMBDA=ALFA
150       ALFA=0.DO
160       FX=FY
          DX=DY
          DO 170 I=1,N
170       X(I)=X(I)+AMBDA*H(I)
          CALL FGR(O)
          FY=F
          DY=0.DO
          DO 180 I=1,N
180       DY=DY+GRATH(I)*H(I)
          IF(DY)190,390,220
190       IF(FY-FX)200,220,220
200       AMBDA=AMBDA+ALFA
          ALFA=AMBDA
          IF(HNRM*AMBDA-1.D10)160,160,210
210       IER=2
          RETURN
220       T=0.DO
230       IF(AMBDA)240,390,240
240       Z=3.DO*(FX-FY)/AMBDA+DX+DY
          ALFA=DMAX1(DABS(Z),DABS(DX),DABS(DY))
          DALFA=Z/ALFA
          DALFA=DALFA*DALFA-DX/ALFA*DY/ALFA
          IF(DALFA)540,250,250
250       W=ALFA*DSQRT(DALFA)
          ALFA=DY-DX+W+W
          IF(ALFA)260,270,260
260       ALFA=(DY-Z+W)/ALFA
          GO TO 280
270       ALFA=(Z+DY-W)/(Z+DX+Z+DY)
280       ALFA=ALFA*AMBDA
          DO 290 I=1,N
290       X(I)=X(I)+(T-ALFA)*H(I)
          CALL FGR(O)
          IF(F-FX)300,300,310
300       IF(F-FY)390,390,310
310       DALFA=0.DO
```

```
              DO 320 I=1,N
320           DALFA=DALFA+GRATH(I)*H(I)
              IF(DALFA)330,360,360
330           IF(F-FX)350,340,360
340           IF(DX-DALFA)350,390,350
350           FX=F
              DX=DALFA
              T=ALFA
              AMBDA=ALFA
              GO TO 230
360           IF(FY-F)380,370,380
370           IF(DY-DALFA)380,390,380
380           FY=F
              DY=DALFA
              AMBDA=AMBDA-ALFA
              GO TO 220
390           IF((OLDF-F)/OLDF-EPS*1.D-20)400,400,400
400           DO 410 J=1,N
              K=N+J
              H(K)=GRATH(J)-H(K)
              K=N+K
410           H(K)=X(J)-H(K)
              IER=0
              IF(KOUNT-N)450,420,420
420           T=0.D0
              Z=0.D0
              DO 430 J=1,N
              K=N+J
              W=H(K)
              K=K+N
              T=T+DABS(H(K))
430           Z=Z+W*H(K)
              IF(HNRM-EPS)440,440,450
440           IF(T-EPS)590,590,450
450           IF(KOUNT-LIMIT)460,530,530
460           ALFA=0.D0
              DO 500 J=1,N
              K=J+N3
              W=0.D0
              DO 490 L=1,N
              KL=N-L
              W=W+H(KL)*H(K)
              IF(L-J)470,480,480
470           K=K+N-L
              GO TO 490
480           K=K+1
490           CONTINUE
              K=N+J
              ALFA=ALFA+W*H(K)
500           H(J)=W
              IF(Z*ALFA)510,10,510
510           K=N31
              DO 520 L=1,N
              KL=N2+L
              DO 520 J=L,N
              NJ=N2+J
              H(K)=H(K)+H(KL)*H(NJ)/Z-H(L)*H(J)/ALFA
520           K=K+1
              GO TO 50
530           IER=1
              RETURN
540           DO 550 J=1,N
              K=N2+J
550           X(J)=H(K)
              CALL FGR(0)
              IF(GNRM-EPS)580,580,560
560           IF(IER.LT.-2)GO TO 590
              IER=IER-1
              GO TO 10
580           IER=0
585           FORMAT(' CONVERGENCE OCCURRED IN FMFP    ')
```

```
            WRITE(LUN,585)
 670        FORMAT(I6,2E20.12)
 590        RETURN
            END


            SUBROUTINE FZEROS
            DOUBLE PRECISION AR(12),AI(12),U,PM,POWER,EXTA(1048)
            COMPLEX RTS(12),ZERO(12),A(13)
            LOGICAL FNREAL
            COMPLEX RT,H,DELFPR,FRTDEF,LAMBDA,DELF,DFPRLM,NUM,DEN,
           +G,SQR,FRT,FRTPRV
            INTEGER KOUNT(12)
            COMMON /BLKO/LX,N,U,PM,POWER/BLK2/EXTA
            EQUIVALENCE (EXTA(289),AR(1)),(EXTA(301),AI(1))
            DATA KN,MAXIT,EP1,EP2,FNREAL/0,40,1.E-12,1.E-20,.FALSE./
            NA=N+1
            LI1=1
            LI2=2
            TPI=ATAN(1.)*8.
            DO 7952 I=1,N
            THH=TPI*RAN(LI1,LI2)
            RTS(I)=1.1*CEXP(CMPLX(0.,THH))
 7952       A(I+1)=CMPLX(SNGL(AR(I)),SNGL(AI(I)))
            A(1)=(1.,0.)
            NAM=NA-1
            EPS1=EP1
            EPS2=EP2
            IBEG=KN+1
            IEND=KN+N
            IND=1
            DO 140 I=IBEG,IEND
            KOUNT(I)=0
 10         H=.5
            RT=RTS(I)+H
            ASSIGN 20 TO NN
            GO TO 60
 20         DELFPR=FRTDEF
            RT=RTS(I)-H
            ASSIGN 30 TO NN
            GO TO 60
 30         FRTPRV=FRTDEF
            DELFPR=FRTPRV-DELFPR
            RT=RTS(I)
            ASSIGN 40 TO NN
            GO TO 60
 40         ASSIGN 110 TO NN
            LAMBDA=-.5
 50         DELF=FRTDEF-FRTPRV
            DFPRLM=DELFPR*LAMBDA
            NUM=-FRTDEF*(1+LAMBDA)*2.
            G=(1+LAMBDA*2.)*DELF-LAMBDA*DFPRLM
            SQR=G*G+2.*NUM*LAMBDA*(DELF-DFPRLM)
            IF(FNREAL.AND.REAL(SQR).LT.0)SQR=0.
            SQR=CSQRT(SQR)
            DEN=G+SQR
            IF(REAL(G)*REAL(SQR)+AIMAG(G)*AIMAG(SQR).LT.0)DEN=G-SQR
            IF(CABS(DEN).EQ.0.)DEN=1.
            LAMBDA=NUM/DEN
            FRTPRV=FRTDEF
            DELFPR=DELF
            H=H*LAMBDA
            RT=RT+H
            IF(KOUNT(I).GT.MAXIT)GO TO 120
 60         KOUNT(I)=KOUNT(I)+1
            FRT=A(NA)
            DO 70 M=1,NAM
 70         FRT=A(NA-M)+RT*FRT
            FRTDEF=FRT
            IF(I.LT.2)GO TO 90
            DO 80 J=2,I
```

```
              DEN=RT-RTS(J-1)
              IF(CABS(DEN).LT.EPS2)GO TO 100
80            FRTDEF=FRTDEF/DEN
90            GO TO NN,(20,30,40,110)
100           RTS(I)=RT+.001
              GO TO 10
110           IF(CABS(H).LT.EPS1*CABS(RT))GO TO 120
              IF(AMAX1(CABS(FRT),CABS(FRTDEF)).LT.EPS2)GO TO 120
              IF(CABS(FRTDEF).LT.10.*CABS(FRTPRV))GO TO 50
              H=H/2
              LAMBDA=LAMBDA/2.
              RT=RT-H
              GO TO 60
120           RTS(I)=RT
              ZERO(I)=FRT
              IF(IND.EQ.0)GO TO 130
              IF(ABS(AIMAG(RTS(I))).LT.1.E-4)GO TO 140
              RTS(I+1)=CONJG(RTS(I))
              IND=0
              GO TO 140
130           IND=1
140           CONTINUE
              WRITE(5,2951)
2951          FORMAT(1X,///,' THE ZEROS OF THE PEF',//,
     +        '       MAG.  & PHASE',//)
              DO 2952 J1=1,N
              RTS(J1)=CONJG(RTS(J1))
              RM=CABS(RTS(J1))
              PHA=ATAN2(AIMAG(RTS(J1)),REAL(RTS(J1)))
              WRITE(5,2953)J1,RM,PHA
2953          FORMAT(3X,I5,E20.7,5X,F20.7,/)
2952          CONTINUE
              RETURN
              END



              DOUBLE PRECISION FUNCTION DFLOAT(I)
              DFLOAT=DBLE(FLOAT(I))
              RETURN
              END



              SUBROUTINE MESG
              DOUBLE PRECISION GR(12,12),GI(12,12),AR(12),AI(12)
     +        ,EXTA(1048),U,PM,POWER
              COMMON /BLK0/LX,LA,U,PM,POWER/BLK2/EXTA
              EQUIVALENCE (EXTA(1),GR(1,1)),(EXTA(145),GI(1,1))
     +        ,(EXTA(289),AR(1)),(EXTA(301),AI(1))
              FPE=(LX+LA+1)/(LX-LA-1)*PM
              WRITE(5,289)
289           FORMAT(1X,///,' LX,LA,U,PM,POWER,FPE',//)
              WRITE(5,251)LX,LA,U,PM,POWER,FPE
251           FORMAT(2X,2I10,4D16.7)
              WRITE(5,991)
991           FORMAT(1X,///,' THE DIAGONAL ELEMENTS OF G',//)
              WRITE(5,299)(GR(LJ,LJ),GI(LJ,LJ),LJ=1,LA)
299           FORMAT(1X,4(D12.5,3X,D12.5,2H ;),/)
              WRITE(5,378)
378           FORMAT(1X,///,' THE PEF COEFS. ',//)
              WRITE(5,299)(AR(LJ),AI(LJ),LJ=1,LA)
              CALL FZEROS
              RETURN
              END
```

REFERENCES

1. **Ahmed, N.,** A Study of Adaptive Digital Filters, Sandia Laboratories Development Rep. 77-0102, Albuquerque, N.M., 1977.
2. **Elliott, G. R., Jacklin, W. L., and Stearns, S. D.,** The Adaptive Digital Filter, SAND 76-0360, Sandia Laboratories, Albuquerque, N.M., 1976.
3. **Elliott, G. R.,** Improving the performance of perimeter security sensors through digital signal processing, in *Proc. Digital Signal Processing Symp.,* SAND 77-1845C, Sandia Laboratories, Albuquerque, N.M., 1977.
4. **Jacklin, W. L.,** Digital Signal Processing Algorithms for the PDP 11, Sandia Laboratories Development Rep. 77-0803, Albuquerque, N.M., 1977.
5. **Ahmed, N.,** On intrusion-detection via adaptive prediction, in *Proc. Digital Signal Processing Symp.,* SAND-77-1845C, Sandia Laboratories, Albuquerque, N.M., 1977.
6. **Anderson, B. D. O. and Moore, J. B.,** *Optimal Filtering,* Prentice-Hall, Englewood Cliffs, N.J., 1979.
7. **Chen, W. Y. and Stegen, G. R.,** Experiments with maximum entropy power spectra of sinusoids, *J. Geophys. Res.,* 79(20), 3019, 1974.
8. **Radoski, H. R., Fougere, P. F., and Zawalick, E. J.,** A comparison of power spectral estimates and applications of the maximum entropy method, *J. Geophys. Res.,* 80(4), 619, 1975.
9. **Fougere, P. F., Zawalick, E. J., and Radoski, H. R.,** Spontaneous line splitting in maximum entropy power spectrum analysis, *Phys. Earth Planet. Inter.,* 12, 201, 1976.
10. **Fougere, P. F.,** A solution to the problem of complex signals, in *Proc. of the RADC Spectrum Estimation Workshop,* Rome Air Development Center, Rome, N.Y., 1978.
11. **Fletcher, R. and Powell, M. J. D.,** A rapid descent method for minimization, *Comput. J.,* 5(2), 163, 1963.
12. **Fletcher, R. and Reeves, C. M.,** Function minimization by conjugate gradients, *Comput. J.,* 7(2), 149, 1964.
13. **Radoski, H. R., Zawalick, E. J., and Fougere, P. F.,** The superiority of maximum entropy power spectrum techniques applied to geomagnetic micropulsations, *Phys. Earth Planet. Inter.,* 12, 208, 1976.
14. **Fougere, P. F.,** Sunspots: power spectra and a forecast, *Int. Solar-Terrestrial Predictions Proc. Workshop Prog.,* Preprint No. 103, National Oceanic and Atmospheric Administration, Boulder, 1979.
15. **Gerhardt, L. A.,** 2nd RADC Spectral Estimation Workshop Problem Competition and Comparison, Rome Air Development Center, Rome, N.Y., 1979.
16. **Ulrych, T. and Bishop, T. N.,** Maximum entropy spectral analysis and autoregressive decomposition, *Rev. Geophys. Space Phys.,* 13, 183, 1975.
17. **Childers, D. G., Ed.,** *Modern Spectrum Analysis,* IEEE, Piscataway, N.J., 1978.
18. **Haykin, S., Ed.,** *Non-linear Methods of Spectral Analysis,* Springer-Verlag, Berlin, 1979.
19. **Chen, C. H.,** Bibliography on Adaptive Digital Filtering and Related Topics, Tech. Rep. prepared for the IEEE Syst. Man and Cybern. Soc., December 1979.

Chapter 9

SONAR SIGNAL PROCESSING

C. H. Chen

TABLE OF CONTENTS

I. INTRODUCTION

A typical treatment of sonar signal processing includes detection theory and techniques, characterization of underwater propagation, active sonar signal processing, passive sonar signal processing, the sonar equation and its uses, and noise measurements, etc. In this chapter, we are concerned only with selected topics in the passive tracking of a moving source, time delay estimation, and an adaptive mean-level detector. These are the areas that particularly require digital processing of sonar waveforms.

II. PASSIVE TRACKING OF A MOVING SOURCE

A signal emitted by a source moving between the ocean surface and the bottom is guided to the observer through direct and specularly reflected ray-paths. The ocean is modeled by a sound velocity whose random part is Gaussian in character with zero mean and whose deterministic part is a constant, c. The water is bounded by two parallel planes separated by a distance h_w. When the region of interest has a layered structure, ray-path studies[1] can be used to describe the wave transmission To formulate a tracking problem, it is highly desirable to choose a simple mathematical model that describes the prominent characteristics of the propagation path. Figure 1 shows the model studied.[2] Assume a receiver to be situated at depth h_{01} and involved in arbitrary motion. It is separated by a distance R from a moving source at depth h_s, as shown in Figure 1. It is desirable to estimate the noisy source motion based on observations of some aspects of this motion conducted at the receiver, as described in Figure 2.

In the ocean context, perhaps the most familiar technique is the two-dimensional tracking of a noisy source by using bearing-only observations. The observer monitors his bearings to a source while proceeding with a constant velocity. He then estimates from these observations its range, course, and speed via a Kalman filter approach. Both moving source (MS) and observer motions are presumed to be in the plane where sequential bearing observations are gathered. Filter convergence is not guaranteed.

For three-dimensional tracking it is well-known in radar-sonar work that serious degradation of elevation or depth measurements is caused by multipath. This happens when the source is at low grazing angles or is within a beam width or so of the bounding surface. A number of techniques have been investigated to reduce multipath errors. The antimultipath techniques are generally ineffective, especially for sources within a beam width of the surface. In shallow water, the difficulty with antimultipath techniques is compounded by the presence of two bounding surfaces.

Figure 2 is designed to track a moving source by a simple observer in shallow water.[2] The observations consist of bearing and time differences between source emitted signals arriving at the receiver via direct, surface, and bottom-reflected paths. The nominal motion is described as the state vector R and its derivative. The Kalman filter formulation can be used under meaningful constraints on the MS motion to obtain an optimal estimate of the source's range, depth, course and speed from the noisy observations.[2]

Processing the received signal yields bearing and time delay observations. The bearing tracker may be some type of correlator, while that of the time delay is an autocorrelator or cepstrum processor. Consider the case of perfect observations: the bearing line is denoted at discrete time k by

$$\beta(k) = \tan^{-1} \{R_x(k)/R_y(k)\} \qquad k = 0,1,2,3,\cdots \qquad (1)$$

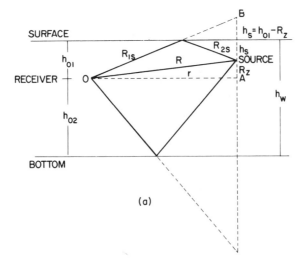

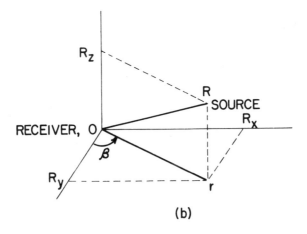

FIGURE 1. Source and receiver geometry.[2] (A) The two-image ray model; (B) coordinate system.

The time delay τ_1 is given by the difference between the surface reflected and direct ray-paths:

$$\tau_1 = (R_{1s} + R_{2s} - R)/C \qquad (2)$$

From the right triangle OAB and $h_s = h_{01} - R_z$, we have

$$R_{1s} + R_{2s} = [r^2 + (h_{01} + h_s)^2]^{1/2}$$
$$= [R^2 + 4 h_{01}^2 - 4 h_{01} R_z]^{1/2} \qquad (3)$$

Thus

$$\tau_1 = ([R^2 + 4 h_{01}^2 - 4 h_{01} R_z]^{1/2} - R)/C \qquad (4)$$

Due to symmetry, the time delay τ_2 given by the difference between the bottom-re-

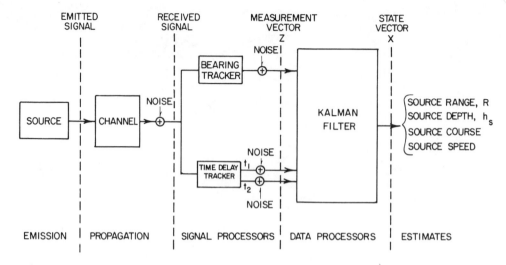

FIGURE 2.   Block diagram of estimation process.[2]

flected and direct paths may be derived from Equation 4 by replacing h_{01} by $-h_{02}$

$$\tau_2 = ([R^2 + 4 h_{01}^2 + 4 h_{02} R_z]^{1/2} - R)/C \tag{5}$$

For long range with $4 h_{01} h_s/R^2 \ll 1$, the time delay τ_1 takes the form

$$\tau_1 = 2 h_{01} (h_{01} - R_z)/RC$$

Similar form can be obtained for τ_2.

With perfect observations, tracking of a moving source is a simple matter. Solutions for the position at the kth observation are easily obtained from Equations 1, 4, and 5.

$$R = \{4(h_{02} h_{01}^2 + h_{01} h_{02}^2) - c^2(\tau_2^2 h_{01} + \tau_1^2 h_{02})\}/2 c(\tau_1 h_{02} + \tau_2 h_{01}) \tag{6}$$

$$R_z = \{c^2\tau_2^2 + 2 c\tau_2 R - 4 h_{02}^2\}/4 h_{02} \tag{7}$$

$$r = (R^2 - R_z^2)^{1/2}, R_x = r \sin \beta, R_y = r \cos \beta \tag{8}$$

With two successive observations, the source velocity may be derived from the above three equations.

In practice the perfect and near perfect observations are seldom available. To account for errors, noise terms n(k) should be included in the mathematical expressions given above. Those errors can be caused by electronic measuring systems of bearing and time delays, by modeling of the environmental factors and source motion, and by monitoring of the observer's own motion.

Now, we consider the state variable formulation. Let $X'(k) = [R_x(k), V_x(k), R_y(k), V_y(k), R_z(k), V_z(h)]$ be a transposed state vector whose components represent the desired variables of source position, $R = [R_x, R_y, R_z]$, and velocity, $V = [V_x, V_y, V_z]$, at a discrete k unit of time, T. The dynamics of the state vector can be modeled by a set of difference equations readily derivable from Newton's laws. The MS acceleration components are modeled by a sample function of a white Gaussian noise process, U(k), with specified covariance, $W(k)$. In a frame of reference attached to the observer, the state equation takes the form

$$X(k+1) = A(k+1/k) \, X(k) - \Gamma(k+1/k) \, V_0(k) + U(k) \qquad (9)$$

where

$$A(k+1/k) = \begin{bmatrix} 1 & T & 0 & 0 & 0 & 0 \\ 0 & 1 & 0 & 0 & 0 & 0 \\ 0 & 0 & 1 & T & 0 & 0 \\ 0 & 0 & 0 & 1 & 0 & 0 \\ 0 & 0 & 0 & 0 & 1 & T \\ 0 & 0 & 0 & 0 & 0 & 1 \end{bmatrix} \qquad (10)$$

and

$$\Gamma(k+1/k) = \begin{bmatrix} T & 0 & 0 \\ 0 & 0 & 0 \\ 0 & T & 0 \\ 0 & 0 & 0 \\ 0 & 0 & T \\ 0 & 0 & 0 \end{bmatrix} \qquad (11)$$

The plant noise $U(k)$ also includes any uncertainty in the description of observer motion denoted by $V_o(k)$.

Observability of MS motion through the measurement vector $Z'(k) = [\beta, \tau_1, \tau_2]$ is indirect. The nonlinear functional relationship between $X(k)$ and $Z(k)$ can be linearized by a Taylor series. The noise should be added to Equations 1, 4, 5. Thus

$$Z(k) = H \, X(k) + n(k) \qquad (12)$$

$$\Delta Z = M \, \Delta \, X + n \qquad (13)$$

and

$$M(k+1 \mid k) = \partial Z / \partial X \big|_{X = X(k+1/k)} \qquad (14)$$

where H is a constant. Kalman filter can now be applied to the observation model. This filter combines each new observation set with preceding MS state estimates and appropriate weighting to yield an improved state estimate.

Given an MS state estimate, $X(k|k)$, and its error covariance, $P(k/k)$, the Kalman filter predicts the MS state, $X(k+1/k)$ at $(k+1)$ along with its predicted covariance $P(k+1|k)$; that is

$$X(k+1 \mid k) = A(k+1 \mid k) \, X(k \mid k) - \Gamma(k+1 \mid k) \, V_0(k) \qquad (15)$$

and

$$P(k+1/k) = A(k+1/k) \, P(k/k) \, A'(k+1/k) + W(k+1) \qquad (16)$$

With an observation $Z(k+1)$, a correction to the predicted values in Equations 15 and

16 can be made by adjusting the filter gain, G(k + 1),

$$G(k+1) = P(k+1/k) \ M'(k+1) \ [M(k+1) \ P(k+1/k) \ M'(k+1) + N(k+1)]^{-1} \qquad (17)$$

The covariance N(k) is that of the noise, n(k). The filtered state vector X(k + 1/k + 1), at (k + 1) and its error covariance P(k + 1/k + 1), are then

$$X(k+1/k+1) \ = \ X(k+1/k) + G(k+1)[Z(k+1) - H \ X(k+1/k)]$$

$$P(k+1/k+1) \ = \ [I - G(k+1) \ M(k+1)] \ P(k+1/k) \qquad (18)$$

and the process is repeated for the next observation. Detailed simulation results and further discussion are available.[2]

III. TIME DELAY ESTIMATION

The time delays referred to in the preceding section are given by the time difference between the surface reflected and direct ray-paths. In this section, the time delays are determined by the time difference between two sensors in receiving a signal from the same source. In general, a knowledge of time delay is useful for estimating the position of a moving source. Consider, for example, two geographically separated sensors that receive a signal from an acoustically radiating point source, as shown in Figure 3. Assume a constant velocity for signal transmission. Then, the travel time from the source to either sensor is directly proportional to the distance traveled. Thus, the difference between travel time from the source to each sensor, or time delay, is given by the difference in path lengths divided by the propagation velocity. The acoustic source must be located on the locus of points that satisfies the constant time delay constraint, namely, the hyberbola in Figure 4. The bearing angle θ that the hyperbolic asymptote makes with the baseline is a good approximation to the true bearing to the source (relative to the midpoint of the baseline), especially for distant sources.

Because the estimation of time delay and bearing is closely related to the coherence between two received waveforms, the coherence function is now briefly described. The complex coherence function is the normalized cross spectral density

$$\gamma_{x_1 x_2}(f) \ = \ \frac{S_{x_1 x_2}(f)}{\sqrt{S_{x_1 x_1}(f) \ S_{x_2 x_2}(f)}} \qquad (19)$$

where $S_{x_1 x_2}(f)$ is the cross spectral density at frequency f between two zero-mean stationary random processes $x_1(t)$ and $x_2(t)$ with auto spectra $S_{x_1 x_1}(f)$ and $S_{x_2 x_2}(f)$. In particular, the magnitude-squared coherence (MSC)

$$|\gamma_{x_1 x_2}(f)|^2 \ = \ C_{x_1 x_2}(f) \ = \ \frac{|S_{x_1 x_2}(f)|^2}{S_{x_1 x_1}(f) \ S_{x_2 x_2}(f)} \qquad (20)$$

has several useful properties[3] and can be effectively estimated by using the fast Fourier transform.[4]

We now consider the maximum likelihood estimate of time delay.[3,5] A signal emanating from a remote source and monitored in the presence of noise at two spatially

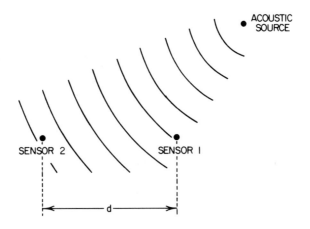

FIGURE 3. Acoustic source and sensors.

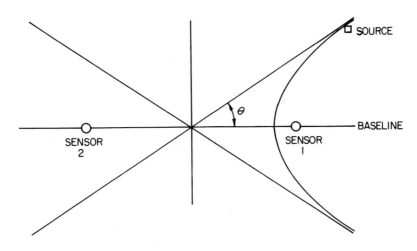

FIGURE 4. Bearing angle interpretation.

separated sensors can be mathematically modeled as

$$x_1(t) = S_1(t) + n_1(t)$$

$$x_2(t) = \alpha S_1(t+D) + n_2(t) \qquad (21)$$

where $S_1(t)$ $n_1(t)$, and $n_2(t)$ are real, jointly stationary random processes. Signal $S_1(t)$ is assumed to be uncorrelated with noises $n_1(t)$ and $n_2(t)$. Also, α is attenuation and D is the time delay to be estimated. The maximum likelihood estimator considered is constrained to operate on observations of a finite duration T.

One common method of determining the time delay and, hence, the arrival angle relative to the sensor axis is to compute the cross-correlation function

$$R_{x_1 x_2}(\tau) = E[x_1(t) x_2(t - \tau)] \qquad (22)$$

where E denotes expectation. The argument τ that maximizes Equation 22 provides an

elay. For ergodic process, an estimate of the cross correlation is given by

$$\hat{R}_{x_1 x_2}(\tau) = \frac{1}{T - \tau} \int_{\tau}^{T} x_1(t)\, x_2(t - \tau)\, dt \tag{23}$$

where T represents the observation interval. In order to improve the accuracy of the delay estimate $\hat{D}$, it is desirable to prefilter $x_1(t)$ and $x_2(t)$ prior to integration in Equation 23. As shown in Figure 5, x_i may be filtered through H_i to yield Y_i for $i = 1,2$. The resultant Y_i are multiplied, integrated, and squared for a range of time shifts τ until the peak is reached which gives a maximum likelihood estimate of the true delay D.

The cross correlation between $x_i(t)$ and $x_2(t)$ is related to the cross spectral density function by the Fourier transform

$$R_{x_1 x_2}(\tau) = \int_{-\infty}^{\infty} S_{x_1 x_2}(f)\, e^{j2\pi f \tau}\, df \tag{24}$$

The cross spectral density of the outputs Y_i is given by[6]

$$S_{y_1 y_2}(f) = H_1(f)\, H_2^*(f)\, S_{x_1 x_2}(f) \tag{25}$$

where * denotes the complex conjugate. A generalized correlation between $Y_1(t)$ and $Y_2(t)$ can be defined as

$$R^{(g)}_{y_1 y_2}(\tau) = \int_{-\infty}^{\infty} W_g(f)\, S_{x_1 x_2}(f)\, e^{j2\pi f \tau}\, df \tag{26}$$

where

$$W_g(f) = H_1(f)\, H_2^*(f) \tag{27}$$

is an appropriately selected weighting function. In practice only an estimate $S_{x_1 x_2}(f)$ of $S_{x_1 x_2}(f)$ can be obtained from finite observations of $x_1(t)$ and $x_2(t)$ Thus, the integral

$$\hat{R}^{(g)}_{y_1 y_2}(\tau) = \int_{-\infty}^{\infty} W_g(f)\, S_{x_1 x_2}(f)\, e^{j2\pi f \tau}\, dt \tag{28}$$

is evaluated and used for estimating delay. The weighting function for the maximum likelihood estimate has been derived as[5,7]

$$W_g(f) = \frac{C_{x_1 x_2}(f)}{|S_{x_1 x_2}(f)|\, [1 - C_{x_1 x_2}(f)]} \tag{29}$$

Other weighting functions have also been considered and compared. Once the time delay is obtained by maximum likelihood estimate, the source bearing can be determined. If the coherence is slowly changing as a function of time, the maximum likelihood estimation of the source bearing will still be a cross correlator preceded by prefilters that must also vary according to time-varying estimate of coherence in a slowly varying environment.

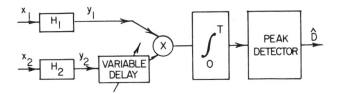

FIGURE 5. Received waveforms filtered, delayed, multiplied, and integrated for a variety of delays until peak output is obtained.

IV. THE ADAPTIVE MEAN-LEVEL DETECTOR

The work reported in this section is motivated by the study of Tuteur.[8] Seismic records may be considered as a special case where the sensors are not moving but the ambient noise is nonstationary. Other applications include the reliable detection of transiting waterborne vessels in the presence of the nonstationary ambient noise of the ocean, as well as similar examples in radar area. The concept of adaptive mean-level detection that makes an estimate of the unknown noise level and uses it in signal detection is very general and should be useful in many applications.

The specific characteristics of ambient noise include

1. Spectral properties.
2. Temporal characteristics. For sonar application, there may be significant variation in seasons, days, or even minutes. For radar application, the atmospheric channel variation is less significant For seismic application, the long propagation path in earth is subject to time variation.
3. Spatial characteristics.
4. Statistical properties.

These characteristics cause a lot of uncertainly in noise level. Figure 6 shows a fixed thresholding system, with threshold set at T. For a constant false alarm rate P_f, the probability of detection P_d decreases as the noise level deviates from design value even when the signal-to-noise ratio remains fixed. Thus, systems whose performance is insensitive to noise level (i.e., only sensitive to signal-to-noise ratio) are much needed for the unknown noise level condition experienced in the transit detection problem. For the seismic records, the transit behavior is due to the fact that the signal level rises and then falls in intensity as the seismic event comes and goes.

Figure 7 shows the functional block diagram of an adaptive mean level threshold system. The low-pass filter provides sufficient smoothing so that x is exponentially distributed[9] or equivalently, chi-square distributed with 2 degrees of freedom in the case when the output of the bandpass filter is stationary. The filter coefficients of the mean estimating filter may be given, but the gain A has to be estimated from the samples at the output of the postdetection filter. The amount of delay and the method of estimating A depend on the application considered. The number of samples useful for estimating A may be small if the noise is highly nonstationary. For sonar application, the performance of the system has been evaluated.[10] Comparison of reverberation level estimates for automatic mean level adjustment has been considered.[11] For seismic discrimination application, the signal to be detected is the explosion or earthquake. The output of the detector is a "noise-free" seismic signal, based on which the explosive and earthquake events can be decided. The estimate of A requires the use of adaptive processing techniques. If space-time processing is required, then predetection combining may be necessary before the mean level detection.

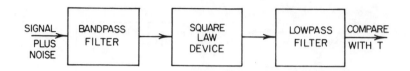

FIGURE 6.   Fixed thresholding system, with threshold set at T.

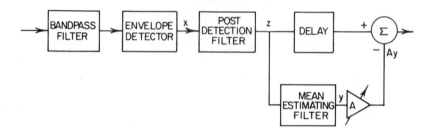

FIGURE 7.   Functional block diagram of an adaptive mean-level threshold system.

BIBLIOGRAPHIC NOTES

On the topic of passively locating an acoustic source, the maximum likelihood method has been proposed[12] to estimate the range and bearing of a source signal observed for a finite time duration at several sensors in the presence of uncorrelated noise. The bearings-only target motion analysis in the ocean context has been examined.[13,14] Cepstrum techniques have been extensively studied for underwater signal extraction, echo detection, and time delay estimation with or without the presence of noise and distortion.[15-21] On the topic of time delay estimation, the efficient estimation of coherence has been considered.[22] Recently, optimum linear filters based on two different criteria were proposed[23] for estimation of time delay by a generalized correlator. The relationships among the various optimum and *ad hoc* filters are also clarified. Pattern recognition methodologies are now playing an increasingly important role in sonar signal processing systems. One problem area[24] is the application and development of suitable recognition algorithms that will identify significant features in a submarine's radiated noise that might make it more detectable or susceptible to being classified and tracked. New methods or new combinations of known methods must be explored to efficiently determine the most significant features for recognition of underwater acoustic signals.

REFERENCES

1. **Officer, C. B.,** *Introduction to the Theory of Sound Transmission,* McGraw-Hill, New York, 1968, chap. 3.
2. **Hassab, J. C.,** Passive tracking of a moving source by a single observer in shallow water, *J. Sound Vib.,* 44(1), 127, 1976.
3. **Carter, G. C.,** Time Delay Estimation, NUSC Tech. Rep. 5335, Naval Underwater Systems Center, New London Laboratory, New London, Conn., 1976.
4. **Carter, G. C., Knapp, C. H., and Nuttall, A. H.,** Estimation of the magnitude-squared coherence function via overlapped fast Fourier transform processing, *IEEE Trans. Audio Electroacoust.,* AU-21(4), 337, 1973.

5. Knapp, C. H. and Carter, G. C., The generalized correlation method for estimation of time delay, *IEEE Trans. Acoustics Speech Signal Process.*, ASSP-24(4), 320, 1976.

6. Davenport, W. B. Jr., *Probability and Random Processes*, McGraw-Hill, New York, 1970.

7. Hannan, E. J. and Thomson, P. J., The estimation of coherence and group delay, *Biometrika*, 58, 469, 1971.

8. Tuteur, F. B., On the detection of transiting broadband targets in noise of uncertain level, *IEEE Trans. Commun. Technol.*, COM-15(1), 61, 1967.

9. Davenport, W. B. and Root, W. L., *Random Signals and Noise*, McGraw-Hill, New York, 1958, 253.

10. Curry, T. J., A Theoretical and Experimental Comparison of Transit Detector Performance in the Time-Varying Noise of the Ocean, Ph.D. thesis, University of Rhode Island, Kingston, 1975.

11. Tufts, D., Unpublished memorandum at the Naval Underwater Systems Center, New London, Conn., June 1975.

12. Carter, G. C., Methods for passively locating an acoustic source, *Records of IEEE Conf. on Acoustics, Speech, and Signal Processing*, IEEE, Piscataway, N.J., 1977.

13. Kolb, R. C. and Hollister, F. H., Bearings-only target motion estimation, *Proc. First Asilomar Conf. on Circuits and Systems*, Pacific Grove, Calif., 1967, 935-946.

14. Murphy, D. J., Noisy Bearings-Only Target Motion Analysis, Ph.D. thesis, Northeastern University, Boston, 1969.

15. Chen, C. H., Digital processing of the marine seismic data, *Proc. IEEE Int. Conf. on Engineering in the Ocean Environment*, IEEE; Piscataway, N.J., 1972.

16. Hassab, J. C. and Boucher, R., Analysis of signal extraction, echo detection and removal by complex cepstrum in presence of distortion and noise, *J. Sound Vib.*, 40, 321, 1975.

17. Hassab, J. C. and Boucher, R., A probabilistic analysis of time delay extraction by the cepstrum in stationary Gaussian noise, *IEEE Trans. Inf. Theory*, It-22(4), 444, 1976.

18. Hassab, J. C. and Boucher, R. E., Further comments on windowing the power cepstrum, *IEEE Trans. Inf. Theory*, 66(10), 1290, 1978.

19. Hassab, J. C., Homomorphic deconvolution is reverberant and distortional channels: an analysis, *J. Sound Vib.*, 58, 215, 1978.

20. Hassab, J. C., The smoothing of "zero" singularities and the effect on time delay detection, *J. Sound Vib.*, 57, 299, 1978.

21. Hassab, J. C. and Boucher, R. E., The effect of dispersive and non-dispersive channels on time delay estimation, *J. Sound Vib.*, 66(2), 247, 1979.

22. Carter, G. C. and Knapp, C. H., Coherence and its estimation via the partitioned modified chirp-Z transform, *IEEE Trans. Acoust. Speech Signal Process.*, ASSP-23(3), 257, 1975.

23. Hassab, J. C. and Boucher, R. E., Optimum estimation of time delay by a generalized correlator, *IEEE Trans. Acoust. Speech Signal Process.*, ASSP-27(4), 373, 1979.

24. Olson, R. C., private communications, 1979.

Chapter 10

APPLICATIONS TO RADAR

C. H. Chen

TABLE OF CONTENTS

I. INTRODUCTION

Digital processing now plays a very important role in radar areas ranging from synthetic aperture radar to clutter suppression, etc. Many solid-state devices have been developed and employed in radar systems to facilitate digital processing. Conventional radar signal processing deals mainly with signal detection, estimation, and design. High-speed spectral computation has added a new dimension to the radar signal processing. In this chapter, selected topics, especially those dealing with the spectral analysis, are considered. Several useful references are available in the 1978 and 1979 Proceedings of the RADC Spectrum Estimation Workshop.

II. DOPPLER SPECTRUM ESTIMATION

The maximum entropy method and other high-resolution spectral analysis methods can significantly increase the signal-to-noise ratio of radar data and thus enhance the signal detection and estimation capability. Conventional Doppler processing uses the discrete Fourier transform. Doppler processing using the maximum entropy spectral estimation has been considered.[1-4] A radar Doppler processor utilizes the effect of Doppler shift on the echo reflected from a moving target in that the power spectrum of this echo is centered about a frequency that is shifted from the transmitted carrier frequency by an amount proportional to the radial velocity of the target. In addition to this target echo, the received signal contains a clutter component, which is produced by reflections from unwanted objects such as ground and weather disturbances, and a received noise component. Figure 1 shows a typical clutter power spectrum density. Here, a signal of small signal-to-noise ratio is located between clutters.

Figure 2 is the block diagram of a Doppler processor[1] using the maximum entropy spectral estimator $\hat{S}_x(f)$, which is designed to compute the spectral estimate of the sampled form of the received signal at a preselected set of frequencies uniformly spaced across the Doppler band of interest. This frequency spacing is determined by the resolution capability of maximum entropy spectral estimate. The logarithm of the spectral estimate is compared with a threshold to determine whether a target is present or not. Due to the nonlinear dependence of $\hat{S}_x(f)$ on the receiver input, computer simulation instead of analytical study was used.[1] The simulation results can be summarized as follows.

In the presence of additive white noise as input, the statistics of the logarithm of the maximum entropy spectral estimate may be closely modeled as Gaussian. The maximum entropy processor is only slightly suboptimal compared to the Doppler processor based on the discrete Fourier transform, which is optimum for the case of a nonfluctuating target. In the presence of both clutter and additive noise, the statistics of the logarithm of the spectral estimate $\hat{S}_x(f)$ deviates from a Gaussian model, with the deviation becoming more pronounced as the slope of the input clutter spectrum is increased. If the clutter component has a narrow spectral width (e.g., ground clutter), the maximum entropy Doppler processor is significantly better than the combination of a double delay-line canceler and discrete Fourier transform for low-Doppler targets.[1,5,6] However, in another experiment which deals with spaceborne, look-down radar,[3] the clutter spectrum shape does not provide any noticeable difference in the results. At least satisfactory detection is achieved by the maximum entropy spectral estimation for any signal location.[4]

In general, a narrow-band Doppler filter is needed to extract targets from clutter. This in turn requires a relatively long sample of the signal. The transmission and reception of this long-time sample occupies the radar time and is not compatible with high search rates and limits radar capability. Reduced dwell times result in lower signal-to-

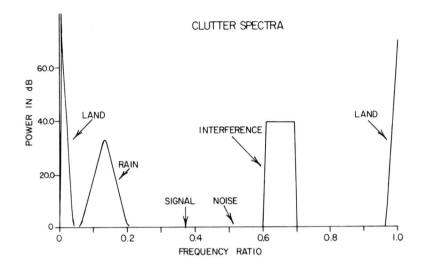

FIGURE 1.   Typical clutter power spectral density.

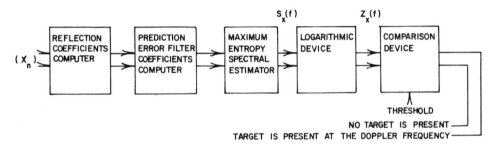

FIGURE 2.   Block diagram of a Doppler processor using the maximum entropy spectral analysis.

interference ratios and poor detection statistics. Thus, the maximum entropy method coupled with fast computation algorithms can meet exactly the need by improving the effective signal-to-interference ratio or the frequency resolution capability of the radar while limiting the dwell time.

At the present stage of development, the maximum entropy Doppler processor does not completely eliminate the need for conventional discrete Fourier transform processing. A maximum entropy based processor may instead considerably extend the performance capability of the conventional radar processors.

III. MULTIPATH PROBLEM IN RADAR TRACKING

The angular resolution and tracking of closely spaced targets is a classical radar problem that is receiving increased attention. It has long been recognized that terrain multipath (e.g., reflections and/or shadowing) are a principal limitation on the achievable accuracy of radar elevation trackers at low elevation angles.[7,8] Since elevation trackers antennas generally have quiet directional patterns in the elevation plane, an important factor in predicting the performance of an elevation tracker is the distribution of the received signal power as a function of elevation angle, i.e., the so-called angular power spectrum.[7] Figure 3 illustrates a typical angular power spectrum that might arise with the multipath environment. The received signal consists of a plane wave at positive elevation angle corresponding to the direct signal coming from the

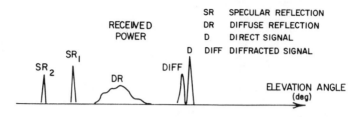

FIGURE 3.   Relationship of received power to low-angle multipath environment.

target, such as an aircraft, and other plane waves, generally at negative elevation angles corresponding to various ground reflections from terrain features. Typically, as shown in Figure 3, the angular power spectrum of the received signal power as a function of elevation angle will consist of a narrow peak at the direct signal elevation angle, narrow peaks at the arrival angles of the major specular ground reflections, and wider peaks in regions of diffuse scattering.[7]

The use of aperture sampling processing to improve the angular resolution/tracking and to characterize the multipath environment has been proposed by Evans and Sun.[9] This approach, which is briefly described in this section, treats the problem as one of estimating the angular power spectrum of the received signal by adaptive processing of the received aperture information. The goals are: (1) to obtain high-resolution, angular power spectra than would be obtainable with the conventional beam sum spectrum for better characterization of multipath environment and (2) to achieve a better estimate of the target elevation angle than would be achievable with the standard tracking methods (e.g., monopulse). To start with, the complex received waveform (i.e., the amplitude and phase) is measured at various points along the receiving antenna aperture (Figure 4). It is noted that there is a duality between angular estimation with line array and frequency estimation for time series. The high-resolution spectral analysis, such as the maximum likelihood and the maximum entropy methods, can then be applied to the spatial data samples.

A radar tracker attempts to determine the centroid of the angular power spectrum corresponding to the direct signal. For an elevation tracker, the direct signal generally corresponds to the peak with the most positive angle. The high-resolution spectral analysis can "sharpen" the spectrum around the direct signal elevation angle and thus reduce the effect of terrain multipaths.

Mathematically, the maximum likelihood estimation of the angular spectrum is given by[9]

$$P_{ML}(\theta) = (e' \hat{R}^{-1} e)^{-1} \tag{1}$$

where

$e = \exp(j2\pi x_k \sin\theta)$ is the received signal vector corresponding to a unit plane wave from angle θ.

$\hat{R}$ is the sample estimate of the covariance matrix $R = E(SS')$ where S represents the complex sensor outputs of the line array and R has as its i, jth entry $R_{ij} = E(S_i'S_j)$

The standard "beam sum" (BS) angle power spectrum estimation is

$$P_{BS}(\theta) = |e'S|^2$$
$$= \left| \sum S(x_k) \exp -j2\pi x_k \sin\theta \right|^2$$
$$= e'\hat{R} e \tag{2}$$

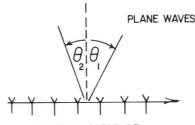

SPATIAL SAMPLE: $S(x_k) = \sum \alpha_i e^{j2\pi \sin\theta_i (x_k)}$

TIME SAMPLE: $S(t_k) = \sum \alpha_i e^{j2\pi f_i t_k}$

x_k = DISPLACEMENT FROM APERTURE PHASE
CENTER IN WAVEL LENGTHS

FIGURE 4.   Linear antenna array and the duality between angular estimation with spatial samples and the estimation for time series.

For the maximum entropy method in radar tracking application, the received angular spectrum can be represented by a finite number of poles in the complex plane, i.e.,

$$P(\theta) = \left| \sum_{i=1}^{N} (\exp j2\pi\delta \sin\theta - Z_i) \right|^{-2} \tag{3}$$

where Z_i lies on or within the unit circle and the spatial samples are taken at points x_k = $k\delta$. The case of Z_i on the unit circle could correspond to discrete plane waves, while Z_i inside the unit circle might correspond to an extended target (e.g., diffuse deflections). Time samples with Equation 3 may be generated by passing a white noise process through an all-pole filter of order N, which is a standard model in autoregressive time series analysis.

Detailed experimental results have been reported for various terrain conditions.[9] It is shown that high-resolution spectral estimation techniques, especially the maximum entropy method, can be effectively used for ground reflection elevation multipath characterization and for improved target elevation angle estimation. Some problems include: (1) the choice of the appropriate prediction error filter length in the maximum entropy method, (2) the proper estimation of the covariance matrix to be used in the maximum likelihood method, and (3) alternative estimators, which are less sensitive to the relative phase between the various received signals.

In closing this section, a note on the use of the maximum entropy method is in order. In an independent study under a different experimental system, Reilly and Haykin[10] employed the maximum entropy method to estimate the wavenumber spectra produced by plane waves, which are incident upon a receiving array as encountered in a low-angle tracking radar. Their experimental results show that, if the direct and reflected components of a composite wave are separated by less than one standard beamwidth, the resulting spectrum could yield false information about the angular separation of these two components unless they are in phase quadrative at the center of the array.

IV. MODELING OF RADAR CLUTTER

Both theoretical and computer models have been examined[11] on the radar clutter.

In the case of clutter generated by reflections of the transmitted wave from various objects located in a radar environment, a popular model of the clutter spectrum is the Gaussian function given by

$$S(f) = S(o) \exp \left(- \frac{f^2}{2\sigma_f^2} \right) \tag{4}$$

where $S(o)$ is the value of the spectral density $S(f)$ at zero frequency and σ_f is the spectral spread. Another proposed model, which shows a better fit of some experimental data, represents $S(f)$ by an all-pole spectrum of order three. Apparently, there is no unique representation of the clutter spectrum that is valid for all circumstances.

The clutter generated in a radar environment may, however, be modeled as an autoregressive (all-pole) process of a finite order (say, an order of three, as mentioned above). On the other hand, the Gaussian-shaped model may be well-approximated by an all-pole model of a relatively low order. For example, the absolute error in approximating the Gaussian function

$$S(f) = \frac{1}{\sqrt{2\pi}} \exp \left(- \frac{f^2}{2} \right)$$

by an autoregressive model of order 6 is 2.7×10^{-3}. Thus, the application of the maximum entropy method to the spectral analysis of radar clutter is theoretically justified. The experimental results reported by Haykin[12] clearly confirm the practical usefulness of Equation 4 and the maximum entropy method.

V. OTHER PROBLEM AREAS

Design of the synthetic aperture radar by using the maximum entropy spectral analysis method has been considered.[13] The method produces narrow mainlobes and negligible sidelobes with very limited data. Problems, however, include longer computing time than the FFT, spectral splitting, and nonlinearity. Further work is needed before the method can be practical for synthetic aperture radar.

Modern pulse Doppler radar systems are required to operate in nonstationary clutter environments, which limit the ability of the radar to detect and track targets if the Doppler filters are not matched to the clutter. An adaptive Doppler filter bank proposed by Sawyers[14] employs the maximum entropy method in deriving the filter coefficients in a nonstationary clutter environment. It is demonstrated by simulation that the adaptive Doppler filters converge rapidly and accurately in several models of clutter and thermal noise.

REFERENCES

1. Haykin, S. and Chan, H. C., The maximum entropy spectral estimator used as a radar Doppler processor, *Proc. 1979 RADC Spectrum Estimation Workshop,* Rome Air Development Center, Rome, N.Y., 1979.

2. Cooper, G. R. and McGillem, C. D., Doppler spectrum estimation for continuously distributed radar targets, *Proc. 1978 RADC Spectrum Estimation Workshop,* Rome Air Development Center, Rome, N.Y., 1979.

3. Tomlinson, P. G. and Ackerson, G. A., Air vehicle detection using advanced spectral estimation techniques, *Proc. 1978 RADC Spectrum Estimation Workshop,* Rome Air Development Center, Rome, N.Y., 1978.

4. King, W. R., Applications for MESA and the prediction error filter, *Proc. 1979 RADC Spectrum Estimation Workshop,* Rome Air Development Center, Rome, N.Y., 1979.

5. McAulay, R. J., A theory for optimal MTI digital signal processing. I. Receiver synthesis, MIT Lincoln Lab. Tech. Note 1972-14, Cambridge, Mass., 1972.

6. Muehe, C. E., Cartledge, L., Drury, W. H., Hofstetter, E.M., Labitt, M., McCorison, P. B., and Sferrino, V. J., New techniques applied to air traffic control radars, *Proc. IEEE,* 62, 716, 1974.

7. Barton, D., Low-angle radar tracking, *Proc. IEEE,* 687, 1974.

8. White, W. D., Low angle radar tracking in the presence of multipath, *IEEE Trans. Aerosp. Electron. Syst.* AES-10(6), 835, 1974.

9. Evans, J. E. and Sun, D. F., Aperture sampling processing for ground reflection elevation multipath characterization, *Proc. 1979 RADC Spectrum Estimation Workshop,* Rome Air Development Center, Rome, N.Y., 1979.

10. Reilly, J. and Haykin, S., An experimental study of the MEM applied to array antennas in the presence of multipath, *Proc. ICASSP,* 80, 120, 1980.

11. Hawkes, C. D. and Haykin, S., Modeling of clutter for coherent pulsed radar, *IEEE Trans. Inf. Theory,* IT-21(6), 703, 1975.

12. Haykin, S., Spectral classification of radar clutter using the maximum entropy method, in *Pattern Recognition and Signal Processing,* Chen, C. H., Ed., Sijthoff & Noordhoff, Winchester, Mass., 1978.

13. Jackson, P. L., Joyce, L. S., and Feldkamp, G. B., Application of maximum entropy frequency analysis to synthetic aperture radar, *Proc. 1978 RADC Spectrum Estimation Workshop,* Rome Air Development Center, Rome, N.Y., 1978.

14. Sawyers, J. H., The maximum entropy method applied to radar adaptive Doppler filtering, *Proc. of 1979 RADC Spectrum Estimation Workshop,* Rome Air Development Center, Rome, N.Y., 1979.

Chapter 11

DIGITAL SYSTEM IMPLEMENTATION

C. H. Chen

TABLE OF CONTENTS

I. INTRODUCTION

The progress in the implementation of digital systems for signal processing and pattern recognition has benefited considerably from the rapid progress in electronic industry on large scale integration and microprocessors, as well as from the significant improvement in mathematical algorithms and software development. In digital signal processing, a major advance is that the fast Fourier transform and the related operations such as convolution and correlation, can now be performed in real time. In pattern recognition, considerable effort was made by industry and government laboratories to implement an automatic recognition system. However, this was met by a lesser amount of success. The recognition accuracy achievable by the machines is usually not comparable with the human performance, thus making a fully automatic system unreliable at this stage of development. The high-speed computation required by digital signal processing is more readily available from the digital computers.

In this chapter, we are concerned with several problem areas in the implementation of digital systems for processing and recognition of waveforms. Many useful details about various aspects of hardware design in signal processing are available.[1-4]

II. THE ARRAY PROCESSOR

The array processor described in Chapter 2 refers to the processing of data from an array of sensors. The array processor discussed here refers to the device and associated software that perform high-speed computation with data sets that are naturally structured in an array or matrix (vector) form. Array processors developed by several digital signal processing companies can now provide extremely fast, affordable computing power, useful for both instrumentation and research. As a peripheral device attached to the host computer, the array processor (AP) handles subroutines. During the execution of the mainline Fortran program, data and instructions are transferred to the AP where the high-speed computations are performed, and the results returned to the host.

The host computer provides overall system control, that is, it controls the flow of data and instructions between I/O (input/output) devices and the array processor. Combining the AP with a general purpose host computer allows each to perform individually in an optimum manner, with the AP taking over the heavy computation and the host computer handling system peripherals, user interaction, and overall control. Combined with minicomputers, the AP has increased the computation speed of systems by factors from 100 to 200. Combined with large computers it provides a computational ability previously available only at tremendous cost. In its simplest implementation, the AP is activated by Fortran calls to AP subroutines. These subroutine calls are automatically handled by the array processor executive, which handles the passing of parameters, setting up any data transfers requested and initiating execution in the AP.

In addition to the host interface, a second path to the APs main data memory is available for direct transfers to and from peripheral devices attached to the AP, such as A/Ds and disks. The disk system provides the needed storage space and high-speed access for applications that have large data bases. Very long programs or large quantities of data can be passed from the host through the AP to its disk storage. The AP can then get instructions and data without interrupting the host.

Typical cycle time of the AP is 167 nsec corresponding to the 6 MHz clock frequency. This is about 60 times faster than the general purpose computer, such as PDP 11/70. Data transfers to or from the host or with peripherals attached to the AP can also take place during computation time. However, time required for transfer is longer than the 167 nsec cycle time. Parallel processing inside the AP itself and pipelining operations can all improve the speed of the processor.

III. EXAMPLE OF FFT BASED SIGNAL PROCESSING SYSTEM

The Fast Fourier Transform (FFT) based signal processor has been widely used in nearly all operational systems that require spectral analysis and digital filtering operations. In this section, we consider an example of a signal processing system that has the speed and versatility to handle a wide variety of underwater acoustic research activities.[5] The design approach is novel in that a fixed-point, highly parallel FFT structure is employed. This approach differs from that reported in radar applications[6, 7] where a cascaded pipeline structure is used. The parallel design is feasible because of significant differences between sonar and radar signal processing requirements. The differences allow the same processor to be used in both forward and inverse transform modes and permit frequency domain techniques to be used in beamforming and signal filtering. The beamforming is to combine various element outputs in the hydrophone array to form directional receiving beams. In passive sonar applications, the signal filtering required is primarily narrowband spectral analysis. Active systems also depend on similar spectral analysis when long carrier wave (CW) tones are transmitted, as in Doppler sonars. In addition, when coded waveforms are used, such as linear frequency modulated (LFM) pulses, the processor must carry out matched filtering or correlation detection.

These basic requirements are common to radar and sonar, but the practical differences are sufficient to support quite different design approaches. For example, the time bandwidth (TW) products, limited to about 200 in sonar, are approximately one order of magnitude less than those employed in radar. Another significant difference is the relatively large Doppler shifts encountered in sonar. Even with relatively Doppler-tolerant signals such as LFM pulses, several replicas must be used in the correlation or matched-filter receiver to cover the anticipated range of target speeds. The fast, parallel structured FFT provides simplification and frequency domain processing freedom to meet the requirements stated above. Beamforming and matched filtering use a number of identical arithmetic operations. Furthermore, beamforming and matched filtering coefficients are simple arrays of complex numbers that can be changed easily to handle different hydrophone arrays and transmitted waveforms. As the data pass completely through the processor in a relatively short time, the parallel structure is much more suitable than the cascaded processor, which has more time delay to get data processed.

Figure 1 is the block diagram of the sonar signal processor.[5] The hydrophone output signals $s_i(t)$ are first sampled and converted to digital form. Up to 30 channels may be handled in real time by this system. Signals from an active system usually arrive in bandpass form and are, therefore, sampled and converted to baseband by quadrature techniques.[8] A relatively high initial sampling rate, e.g., 5 kHz per channel, is used. This is large compared to the signal bandwidths but permits the use of simple analog band-limiting filters. A sharp cutoff linear-phase digital filter then allows the sampling rate to be reduced to a more efficient value, on the order of 500 to 1000 complex samples per second, without aliasing.

The outputs of the digital filter are stored in a serial access memory (SAM) to await transformation to the frequency domain by the FFT unit. The transformed data are stored in a "corner turning" memory in a serial form as a function of frequency. During beamforming the data are processed in the order of the hydrophone elements. The array processor (AP) is essentially a complex multiplier and accumulator that per-

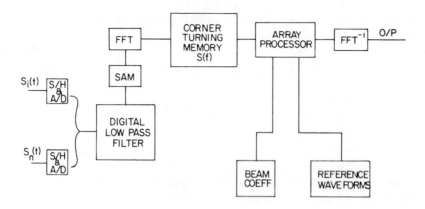

FIGURE 1.   Block diagram of a sonar signal processor.[5]

forms beamforming and matched filtering. In the beamforming mode, AP draws each hydrophone signal component from the corner-turning memory, multiplies it by appropriate complex weighting coefficients, and accumulates or sums the products. This operation is repeated for each frequency and for each beam. When completed, data that were originally in element-frequency order are rewritten as direction-frequency data in the corner-turning memory.

In the matched filtering mode for coded signals, the frequency components from each beam are multiplied by complex weighting coefficients representing the Fourier transform of the replica waveform. The resulting sequence is inverse Fourier transformed (FFT⁻¹) to obtain the cross correlation as a function of time delay between the signal and the reference waveform. After the matched filtering operation is completed, the corner-turning memory contains its final result, which is the decomposition of the signal from the hydrophone array into range (time delay) and bearing (direction) components.

Now we can examine in more detail the demands of beamforming and filtering and the way they are handled by the FFT processor. To do the beamforming in the frequency domain, corresponding spectral components from a number of hydrophone elements are selected from the corner-turning memory, multiplied in the array processor with appropriate phase and amplitude shading coefficients and summed. Sufficient time for the array processor to carry out these calculations is provided by processing the input data through the FFT as high-speed bursts. This is really only practical with the parallel FFT structure, which can complete spectral analysis on ten data channels within the time required just for start-up of a serial FFT. For matched filtering of coded pulses, multiple reference waveforms required to handle echoes with relatively large Doppler shifts in sonar application again make it necessary to use the parallel FFT implementation to process bursts of data at mixed data rates and to handle a mixture of forward and inverse transforms. Fixed point scaling[9] can be used in matched filtering to prevent overflow and to overcome nonlinearity in the signal flow.

It should be noted that alternative Fourier transform and convolution methods based on the number theory are now available.[10] The major development includes the Winograd Fourier Transform algorithm and the number-theoretic transforms. These transform algorithms promise to provide significant improvement over the existing FFT algorithms. For example, the result of the hardware comparison[11] of the FNT (Fermat Number Transform) vs. a pipeline FFT indicates that cost savings of the FNT can be realized for small systems (e.g., length 64 convolution) when the signal to be filtered is real.

IV. PATTERN RECOGNITION HARDWARE

For digital waveforms, speech processing and recognition system implementation has been most successful (Chapter 6). Processing and recognition are integrated into one system in most applications. Most efforts on pattern recognition hardware, however, are for imagery patterns that may involve optical devices. Many systems developed are general purpose in application. For example, flexible templates (rubber masks) have been devised for use in a complete pattern measurement, analysis, and recognition system[12] useful for EEG study and other applications. Interactive pattern analysis and recognition systems, such as that developed at Rome Air Development Center, depend more on software than on hardware to perform recognition tasks. Among other developments, charge-coupled devices[13] and microprocessors have been considered for intrusion-detection system hardware using adaptive digital filtering.

V. CONCLUDING REMARKS

The last decade has seen many developments in LSI (Large Scale Integration) and VLSI (Very Large Scale Integration) technologies. These developments have already had profound impact on digital signal processing.[14] In the future, considerable additional performance improvement may be expected, primarily from size reductions with lower cost and operating power as well as increasing speed. The next decade should see a significant increase in the application of digital signal processing, with extensions into areas now considered impractical. The digital system implementation problem will be increasingly important as every new algorithm developed must be suitable for practical implementation by hardware. Recognition systems that require digital signal processing for preprocessing and for feature extraction will also benefit from advances in the electronic industry. Presently, there is far more software than hardware effort in pattern recognition. With the increased demand for automatic waveform recognition, much progress in recognition hardware is anticipated in the next decade.[15-19]

REFERENCES

1. IEEE NEREM Record on Signal Processing, IEEE, Boston, 1973.
2. Oppenheim, A. V., Ed., *Applications of Digital Signal Processing,* Prentice Hall, Englewood Cliffs, N.J., 1978.
3. Rabiner, L. R. and Gold, B., *Theory and Application of Digital Signal Processing,* Prentice-Hall, Englewood Cliffs, N.J., 1975.
4. Peled, A. and Liu, B., *Digital Signal Processing, Theory, Design and Implementation,* John Wiley & Sons, New York, 1976.
5. Trider, R. C., A fast Fourier transform (FFT) based sonar signal processor, *IEEE Trans. Acoust. Speech Signal Process.,* ASSP-26(1), 15, 1978.
6. Blankenship, P. E. and Hofstetter, E. M., Digital pulse compression via fast convolution, *IEEE Trans. Acoust. Speech Signal Process.,* ASSP-23, 189, 1975.
7. Martinson, L. W. and Smith, R. J., Digital matched filtering with pipelined floating point fast Fourier transform (FFT's), *IEEE Trans. Acoust. Speech Signal Process.,* ASSP-23, 222, 1975.
8. Sears, R. W., Jr. and Talpey, T. E., A narrowband multichannel sampled data acquisition and processing system for underwater acoustic measurements, *IEEE Trans. Audio Electroacoust.,* AU-19(2), 174, 1971.
9. Oppenheim, A. V. and Schafter, R. W., *Digital Signal Processing,* Prentice-Hall, Englewood Cliffs, N.J., 1975.
10. McClellan, J. H. and Rader, C. M., *Number Theory in Digital Signal Processing,* Prentice-Hall, Englewood Cliffs, N.J., 1979.

11. **McClellan, J. H.,** Hardware realization of a Fermat Number Transform, *IEEE Trans. Acoust. Speech Signal Process.,* ASSP-24(3), 189, 1976.

12. **Widrow, B.,** The 'rubber mask' technique. I. Pattern measurement and analysis. II. Pattern storage and recognition, *Pattern Recognition,* 5, 175, 1973.

13. **Donohoe, G. W.,** A hardware implementation of adaptive filtering using charge-coupled devices, *Proc. Digital Signal Processing Symp.,* Sandia Laboratories, Albuquerque, N.M., 1977.

14. **Hoff, M. E., Jr.,** IC Technology: Trends and impact on digital signal processing, *Proc. ICASSP,* 80, 1, 1980.

15. **Blesser, B.,** Digital processing in audio signals, in *Applications of Digital Signal Processing,* Oppenheim, A. V., Ed., Prentice-Hall, Englewood Cliffs, N.J., 1978.

16. **Swartzlander, E. E.,** Signal processing architectures with VLSI, *Proc. ICASSP,* 80, 368, 1980.

17. **Thompson, J. S. and Boddie, J. R.,** An LSI digital signal processor, *Proc. ICASSP,* 80, 383, 1980.*

18. **Edwards, G. P.,** A speech/speaker recognition and response system, *Proc. ICASSP,* 80, 394, 1980.

19. **McCormick, B. H.,** The Illinois pattern recognition computer, *IEEE Trans. Electron. Comput.,* EC-12, 791, 1963.

* This processor is designed particularly for telecommunications applications.

BIBLIOGRAPHY

1. **Ahmed, N., Hummels, D. R., Uhl, M., and Soldan, D.,** A short-term sequential regression algorithm, *IEEE Trans. Acoust. Speech Signal Process.,* 27, 453, 1979.
2. **Anuta, P. E.,** Signal Processing and Analysis of Combined Data Types for Geophysical Exploration, Workshop on Digital Signal and Waveform Analysis, Miami Beach, December 5, 1980.
3. **Avgeris, T., Lithopoulos, E., and Tzannes, N. S.,** Application of the mutual information principle to spectral density estimation, *IEEE Trans. Inf. Theory,* IT-26(2), 184, 1980.
4. **Aytun, K. and Gülünay, N.,** Computation of Instantaneous Parameters of Seismic Signal with Applications to Black Sea and Mediterranean Data, paper presented at the 26th Congr. of CIESM, Antalya, Turkey, December 1978.
5. **Ben-Bassat, M., Carlson, R. W., Puri, V. K., Davenport, M. D., Schriver, J. A., Latif, M., Smith, R., Portigal, L. D., Lipnick, E. H., and Weil, M. H.,** Pattern-based interactive diagnosis of multiple disorders: the MEDAS system, *IEEE Trans. Pattern Anal. Mach. Intelligence,* PAMI-2(2), 148, 1980.
6. **Berni, A. J.,** Target identification by natural resonance estimation, *IEEE Trans. Aerosp. Electron. Syst.,* AES-11(2), 147, 1975.
7. **Bourne, J. R.,** Syntactic Analysis of Electroencephalographic Signals, Workshop on Digital Signal and Waveform Analysis, Miami Beach, December 5, 1980.
8. **Carter, G. C.,** Bias in magnitude-squared coherence estimation due to misalignment, *IEEE Trans. Acoust. Speech Signal Process.,* 28(1), 97, 1980.
9. **Chan, Y. T., Riley, J. M., and Plant, J. B.,** A parameter estimation approach to time-delay estimation and signal detection, *IEEE Trans. Acoust. Speech Signal Process.,* 28(1), 8, 1980.
10. **Chen, C. H.,** Adaptive and Learning Algorithms for Intrusion-Detection with Seismic Sensor Data, Workshop on Digital Signal and Waveform Analysis, Miami Beach, December 5, 1980.
11. **Chittineni, C. B.,** Signal Classification for Automatic Industrial Inspection, Workshop on Digital Signal and Waveform Analysis, Miami Beach, December 5, 1980.
12. **Clark, B. L.,** A comparative evaluation of several bearings-only tracking filters, *Proc. ICASSP,* 3, 833, 1980.
13. **Coker, M. J. and Simkins, D. N.,** A nonlinear adaptive noise canceller, *Proc. ICASSP,* 80(2), 470, 1980.
14. **Dasarathy, B. V. and Sheela, B. V.,** A composite classifier system design: concepts and methodology, *Proc. IEEE,* 67(5), 708, 1979.
15. **Davis, D. L. and Bouldin, D. W.,** A cluster separation measure, *IEEE Trans. Pattern Anal. Mach. Intelligence,* PAMI-1 (2), 224, 1979.
16. **Dove, W. P. and Oppenheim, A. V.,** Event location using recursive least squares signal processing, *Proc. ICASSP,* 3, 848, 1980.
17. **deFigueiredo, R. J. P.,** Application of a Frequency Domain Prony Method to Wide Bandwidth Radar Signature Classification, Workshop on Digital Signal and Waveform Analysis, Miami Beach, December 5, 1980.
18. **Dwyer, R. F.,** Detection of partitioned signals by discrete cross-spectrum analysis, *Proc. ICASSP,* 2, 638, 1980.
19. **Farag, R. F. H. and Rothweiler, J. H.,** Signal Processing and Pattern Recognition for Intelligent Sensors, Workshop on Digital Signal and Waveform Analysis, Miami Beach, December 5, 1980.
20. **Flora, J. H., Smith, S., and Jain, A. K.,** Computer Classification of Eddy Current Signals for Automatic Inspection of Steam Generator Tubes, Workshop on Digital Signal and Waveform Analysis, Miami Beach, December 5, 1980.
21. **Fougere, P. F.,** Sunspots: power spectra and a forecast, *Int. Solar-Terrestrial Predictions Proc. Workshop Prog.,* Preprint No. 103, National Oceanic and Atmospheric Association, Boulder, 1979.
22. **Fu, K. S.,** On mixed approaches to pattern recognition, *Proc. IEEE Int. Conf. on Cybern. Soc.,* Piscataway, N.J., 1980, 930.
23. **Fukunaga, K. and Short, R. D.,** A class of feature extraction criteria and its relation to the Bayes risk estimate, *IEEE Trans. Inf. Theory,* IT-26(1), 59, 1980.
24. **Gerhardt, L. A.,** The Use and Comparison of Spectral Estimators for Signal Processing and Detection with Applications, Workshop on Digital Signal and Waveform Analysis, Miami Beach, December 5, 1980.
25. **Gibson, C. and Haykin, S.,** Adaptive Learning Characteristics of Maximum Entropy (Lattice) Filtering Algorithms , *IEEE Trans. Acoust. Speech Signal Process,* in press.
26. **Griffiths, L. J., Smolka, F. R., and Trembly, L. D.,** Adaptive deconvolution: a new technique for processing time-varying seismic data, *Geophysics,* 42(4), 742, 1977.
27. **Guarino, C. R.,** Tone Estimation Using an Autoregressive-Moving Average Model, Workshop on Digital Signal and Waveform Analysis, Miami Beach, December, 5, 1980.

28. Hassab, J. C., Guimond, B. W., and Nardone, S. C., *Proc. ICASSP*, 3, 811, 1980.
29. Herring, R. W., The cause of line-splitting in Burg maximum entropy spectral analysis, *IEEE Trans. Acoust. Speech Signal Process.*, ASSP-28(6), 692, 1980.
30. Jain, V. K., Efficient Signal Representation and Approximation, Workshop on Digital Signal and Waveform Analysis, Miami Beach, December 5, 1980.
31. Kaiser, J. F. and Schafter, R. W., On the use of the I_o-Sinh window for spectrum analysis, *IEEE Trans. Acoust. Speech Signal Process.*, 28(1), 105, 1980.
32. Kanal, L. N. and Stockman, G., On a Problem Reduction Approach to the Linguistic Analysis of Waveforms, 1980 Workshop on Digital Signal and Waveform Analysis, Miami Beach, Fla., December 5, 1980.
33. Kazakos, D. and Papantoni-Kazakos, P., Spectral distance measures between Gaussian processes, *Proc. ICASSP*, 2, 612, 1980.
34. Lang, S. W., Near optimal frequency/angle of arrival estimates based on maximum entropy spectral techniques, *Proc. ICASSP*, 3, 829, 1980.
35. Lin, Y. K., The Seismological Pattern Analysis and Classification with ACDA Data Base, AARL Memo No. 29, Purdue University, West Lafayette, Ind., January 1978.
36. Liu, S. C. and Nolte, L. W., Performance evaluation of array processors for detecting Gaussian acoustic signals, *IEEE Trans. Acoust. Speech Signal Process.*, 28(3), 328, 1980.
37. Makhoul, J., A fast cosine transform in one and two dimensions, *IEEE Trans. Acoust. Speech Signal Process.*, 28(1), 27, 1980.
38. Messerschmitt, D. G., A class of generalized lattice filters, *IEEE Trans. Acoust. Speech Signal Process.*, 28(2), 198, 1980.
39. Mick, J. R. and New, B. J., Bit slice devices for signal processing, *Proc. ICASSP*, 80(2), 372, 1980.
40. Mintzer, F., Attributes of parallel and cascade microprocessor implementations of digital signal processing, *Proc. ICASSP*, 3, 912, 1980.
41. Moore, J. D. and Lawrence, N. B., Signal-to-noise losses associated with hardware-efficient designs of moving target indicators, *IEEE Trans. Acoust. Speech Signal Process.*, 28(1), 35, 1980.
42. Morgan, D. R. and Craig, S. E., Real-time adaptive linear predication using the least mean square gradient algorithm, *IEEE Trans. Acoust. Speech Signal Process.*, 24(6), 494, 1976.
43. Mottl', V. V. and Muchnik, I. B., Linguistic analysis of experimental curves, *Proc. IEEE*, 67(5), 714, 1979.
44. Munson, D. C., Jr. and Liu, B., Low noise realizations for narrow-band recursive digital filters, *IEEE Trans. Acoust. Speech Signal Process.*, 28(1), 41, 1980.
45. Nuttal A. H. and Carter, G. C., A generalized framework for power spectral estimation, *IEEE Trans. Acoust. Speech Signal Process.*, 28(3), 334, 1980.
46. Pau, L. F., An adaptive signal classification procedure: application to aircraft engine condition monitoring, *Pattern Recognition*, 9(3), 121, 1977.
47. Portnoff, M. R., Time frequency representation of digital signals and systems based on short-time Fourier analysis, *IEEE Trans. Acoust. Speech Signal Process.*, 28(1), 8, 1980.
48. Rabiner, L. R. and Allen, J. B., On the implementation of a short-time spectral analysis method for system identification, *IEEE Trans. Acoust. Speech Signal Process.*, 28(1), 69, 1980.
49. Rabiner, L. R. and Schafter, R. W., *Digital Processing of Speech Signals*, Prentice-Hall, Englewood Cliffs, N.J., 1978.
50. Riccia, G. D., Feature processing by optimal factor analysis techniques in statistical pattern recognition, *IEEE Trans. Pattern Anal. Mach. Intelligence*, PAMI-2(1), 60, 1980.
51. Robinson, E. A. and Treitel, S., Maximum entropy and the relationships of the partial autocorrelation to the reflection coefficients of a layered systems, *IEEE Trans. Acoust. Speech Signal Process.*, 28(2), 224, 1980.
52. Sankar, P. V. and Rosenfeld, A., Hierarchical representation of waveforms, *IEEE Trans. Pattern Anal. Mach. Intelligence*, PAMI-1(1), 73, 1979.
53. Scharf, L. L. and Luby, J. C., Statistical design of autoregressive-moving average digital filters, *IEEE Trans. Acoust. Speech Signal Process.*, 27(3), 240, 1979.
54. Schiess, J. R. Zero Crossing Counts as a Method of Classifying Digital Signals, paper presented at the Joint Statistical Meetings, Washington, D.C., August 1979.
55. Siegel, L. J. and Bessey, A. C., A decision tree procedure for voiced/unvoiced/mixed excitation classification of speech, *Proc. ICASSP*, 80(1), 53, 1980.
56. Silvia, M. T., Deconvolution of Geophysical Time Series, Ph.D. thesis, Northeastern University, Boston, Mass., 1977.
57. Sklansky, J. and Michelotti, L., Locally trained piecewise linear classifiers, *IEEE Trans. Pattern Anal. Mach. Intelligence*, PAMI-2(2), 101, 1980.
58. Smith, A. F. M. and Makov, U. E., A quasi-Bayes sequential procedure for mixtures, *J. R. Stat. Soc. Ser. B*, 40, 106, 1978.

59. **Srinath, M. D. and Rajasekaran, P. K.,** An Introduction to Statistical Signal Processing with Applications, John Wiley & Sons, New York, 1979.

60. **Steiglitz, K.,** On the simultaneous estimation of poles and zeros in speech analysis, *IEEE Trans. Acoust. Speech Signal Process.* 25, 229, 1977.

61. **Swingler, D. N.,** A comparison between Burg's method and nonrecursive technique for maximum entropy processing of deterministic signals, *J. Geophys. Res.,* 84(B2), 679, 1979.

62. **Therrien, C. W.,** A sequential approach to target discrimination, *IEEE Trans. Aerosp. Electron. Syst.,* AES-14(3), 433, 1978.

63. **Tominaga, S.,** Analysis of Experimental Curves Using Singular Value Decomposition, *IEEE Trans. Acoust. Speech Signal Process.,* in press.

64. **Treichler, J. R.,** Response of the adaptive line enhancer to chirped and doppler-shifted sinusoids, *IEEE Trans. Acoust. Speech Signal Process.,* 28(3), 343, 1980.

65. **Ursin, B.,** Seismic signal detection and parameter estimation, *Geophys. Prospect.,* 27, 1, 1980.

66. **Wallach, Y. and Shimor, A.,** Alternating sequential-parallel versions of FFT, *IEEE Trans. Acoust. Speech Signal Process.,* 28(2), 236, 1980.

67. **Walton, E. K.,** The Use of Phasor Space Representation of Radar Data in a Radar Target Identification System, Workshop on Digital Signal and Waveform Analysis, Miami Beach, December 5, 1980.

68. **Waser, S.,** Survey of VLSI for digital signal processing, *Proc. ICASSP,* 80(2), 376, 1980.

69. **Wegman, E. J.,** Optimal estimation of time series functions, *IEEE Trans. Acoust. Speech Signal Process.,* ASSP-28(6), 763, 1980.

70. **Wegman, E. J.,** Two approaches to nonparametric regression: splines and isotonic inference, in *Recent Development in Statistical Inference and Data Analysis,* Matusita, K., Ed., North-Holland, New York, in press.

71. **Willsky, A. S.,** *Digital Signal Processing and Control and Estimation Theory,* MIT Press, Cambridge, 1979.

72. **Wong, A. C. and Wang, D. C. C.,** DECA: a discrete-valued data clustering algorithm, *IEEE Trans. Pattern Anal. Mach. Intelligence,* PAMI-1(4), 342, 1979.

73. **Yablon, M. and Chu, J. T.,** The relationship of the Bayes risk to certain separability measure in normal classification, *IEEE Trans. Pattern Anal. Mach. Intelligence,* PAMI-2(2), 97, 1980.

INDEX

D

G

H

I

K

L

N

O

P